Recent Advances in
BIOCHEMISTRY

Recent Advances in
BIOCHEMISTRY

Harold H. Trimm, PhD, RSO

Chairman, Chemistry Department, Broome Community College;
Adjunct Analytical Professor, Binghamton University,
Binghamton, New York, U.S.A.

William Hunter III

Researcher, National Science Foundation, U.S.A.

Apple Academic Press

Recent Advances in Biochemistry

© Copyright 2011*
Apple Academic Press Inc.

This book contains information obtained from authentic and highly regarded sources. A wide variety of references are listed. Reasonable efforts have been made to publish reliable data and information, but the editors and the publisher cannot assume responsibility for the validity of all materials or for the consequences of their use.

First Published in the Canada, 2011
Apple Academic Press Inc.
3333 Mistwell Crescent
Oakville, ON L6L 0A2
Tel. : (888) 241-2035
Fax: (866) 222-9549
E-mail: info@appleacademicpress.com
www.appleacademicpress.com

The full-color tables, figures, diagrams, and images in this book may be viewed at www.appleacademicpress.com

First issued in paperback 2021

ISBN 13: 978-1-77463-257-4 (pbk)
ISBN 13: 978-1-926692-72-2 (hbk)

Harold H. Trimm, PhD, RSO
William Hunter III

Cover Design: Psqua

Library and Archives Canada Cataloguing in Publication Data
CIP Data on file with the Library and Archives Canada

To the best of the publisher's knowledge, all articles in this book are copyrighted to the individual authors and are licensed and distributed under the Creative Commons Attribution License, which permits unrestricted use, distribution, and reproduction in any medium, provided the original work is properly cited.

CONTENTS

INTRODUCTION

Biochemistry is the study of the chemistry of living systems, such as plants and animals. Currently, most research falls into three main categories, plant, general, and human. Biochemistry is closely allied to the fields of genetics and molecular biology. Biochemists study the structure, function, and synthesis of molecules involved in the living process. Advances in biochemistry directly influence medicine and the field of human health.

The structural unit of life is the cell. Living organisms can consist of a single cell, such as phytoplankton or bacteria, or can be made up of a complex organization of 100 trillion cells, as humans are. Cells that contain chloroplasts are classified as plants and can use sunlight to create chemical energy stored as carbohydrates through the process of photosynthesis. Cells also contain ribosomes, where amino acids are assembled into proteins that can be structural units, such hemoglobin, or enzymes, such as amylase, which catalyze specific reactions in the cell. Mitochondria, also contained in cells, are the powerhouse of the cell, providing the energy needed for the reactions of life.

Biochemistry plays a central role in all living things. It studies the energy flow from molecules such as sugars and carbohydrates, the catalysis of cellular reaction by enzymes, the role of nonpolar constituents such as lipids and fatty acids, structural components such as proteins and biopolymers, and inheritance based on DNA.

Modern advances in biochemistry include the 2008 Nobel Prize in chemistry awarded to the researchers of green fluorescent protein (GFP). This glowing jellyfish protein has been manipulated by biochemists to revolutionize the tracking of tumors and cancer cells. In the area of alternative fuels, biochemists have been developing enzymes to break down the cellulose of wood to produce ethanol by a process more efficient than corn fermentation. Other groups of biochemists are researching the use of algae to sequester carbon dioxide from various sources and use the oil produced by the algae for biodiesel. Biochemists are constantly adding to our knowledge of what reactions are involved in all living systems.

Much of the recent advances in biochemistry are based on the availability of new instrumentation. The limitations of older techniques prevented their use in studying high molecular weight components involved in biological systems. The growing availability of liquid chromatography mass spectrometers (LC-MS), which can analyze high molecular weight biomolecules, has led to advances in pharmacokinetics, proteology, and drug development. The use of inductively coupled plasma mass spectrometers (ICP-MS) is now being adopted by biochemists to study protein and biomolecules speciation.

Some of the areas where biochemists are employed include agriculture, biotechnology, cancer research, chemical manufacturing, environmental pollution control, food and drink, forensics, hazardous waste, hospitals, pharmaceuticals, petroleum, polymer, paper, public health, research, universities, and water treatment. The largest employer of biochemists is the pharmaceutical industry.

The field of biochemistry is constantly changing, with new discoveries being made all the time. There is an increased focus on green chemistry, which is the design of chemical products and processes that reduce or eliminate the use or generation of hazardous substances. In the medical field there is new research being done on detecting disease on a molecular and genetic level and designing new drugs to cure these diseases. There is even research into identifying genetic markers to predict diseases before they occur. Biofuels are another growing research area. The conversion of natural products into energy sources is of vital importance to the future.

— **Harold H. Trimm, PhD, RSO**

Biochemical Characterization of Bovine Plasma Thrombin-Activatable Fibrinolysis Inhibitor (TAFI)

Zuzana Valnickova, Morten Thaysen-Andersen, Peter Højrup, Trine Christensen, Kristian W. Sanggaard, Torsten Kristensen and Jan J. Enghild

ABSTRACT

Background

TAFI is a plasma protein assumed to be an important link between coagulation and fibrinolysis. The three-dimensional crystal structures of authentic mature bovine TAFI (TAFIa) in complex with tick carboxypeptidase inhibitor, authentic full length bovine plasma thrombin-activatable fibrinolysis inhibitor (TAFI), and recombinant human TAFI have recently been solved. In light of these recent advances, we have characterized authentic bovine TAFI biochemically and compared it to human TAFI.

Results

The four N-linked glycosylation sequons within the activation peptide were all occupied in bovine TAFI, similar to human TAFI, while the sequon located within the enzyme moiety of the bovine protein was non-glycosylated. The enzymatic stability and the kinetic constants of TAFIa differed somewhat between the two proteins, as did the isoelectric point of TAFI, but not TAFIa. Equivalent to human TAFI, bovine TAFI was a substrate for transglutaminases and could be proteolytically cleaved by trypsin or thrombin/solulin complex, although small differences in the fragmentation patterns were observed. Furthermore, bovine TAFI exhibited intrinsic activity and TAFIa attenuated tPA-mediated fibrinolysis similar to the human protein.

Conclusion

The findings presented here suggest that the properties of these two orthologous proteins are similar and that conclusions reached using the bovine TAFI may be extrapolated to the human protein.

Background

Human thrombin-activatable fibrinolysis inhibitor (TAFI) (EC 3.4.17.20; Uni-Prot, Q96IY4), also known as plasma pro-carboxypeptidase B, R, and U, is a plasma metallocarboxypeptidase that attenuates fibrinolysis [1-10]. TAFI circulates in plasma as a 58 kDa protein with significant intrinsic activity [11,12]. The majority of the sites that undergo transglutaminase-mediated cross-linking to fibrin are primarily located on the heavily glycosylated pro-peptide, suggesting that TAFI becomes incorporated into the fibrin clot during later stages of the coagulation cascade [13]. A variety of trypsin-like proteinases have been shown to remove this peptide, generating the mature protein, TAFIa [4,14-17]. The isoelectric point (pI) of this proteolytically cleaved protein is around pH 8.5, which is significantly more basic than that of TAFI (pI 5.5) [18]. TAFIa remains in circulation by forming complexes with α_2-macroglobulin and pregnancy zone protein [19] but is highly unstable, a feature initially attributed to proteolytic cleavage. However, this instability is now thought to result from a temperature-dependent conformational change that occurs within minutes of activation [4,20-22].

TAFI has been implicated not only in fibrinolysis, but also in inflammation, wound healing, and a variety of other deficiencies and diseases, such as diabetes, kidney failure, lung cancer, and liver illnesses [23-29]. Interestingly, individuals with the more stable Ile325 variant are apparently more susceptible to meningococcal sepsis [30]. TAFI has been studied in multiple animal models, including dog, rabbit, mouse, and rat [31-36]. Intriguingly, the absence of the protein in knock out mice is compatible with murine life [25,37,38].

Mouse and rat TAFI have been characterized, and both show similarity to the human protein [32,33,35]. Until very recently, the only available structural model for the study of TAFI was human pancreatic pro carboxypeptidase B (pro-CPB) [39]. The protein sequence of Pro-CPB is about 40% identical to TAFI. However, in contrast to TAFI, pro-CPB lacks intrinsic activity and its active form, carboxypeptidase B (CPB), is stable upon activation [40]. Efforts to crystallize authentic human TAFI have been unsuccessful, most likely due to its sugar heterogeneity when purified from pooled plasma [18]. However, using recombinant human TAFI and authentic protein purified from a single cow enabled the zymogen structure to be solved [41,42]. Although bovine TAFI is similar to pro-CPB, it also has differences. Significantly, the position of the pro-peptide is rotated 12° away from the active site, exposing access to the catalytic residues. Another significant distinction is the lack of the corresponding salt bridge between Asp41 and Arg145 in TAFI [42]. These distinctions might explain the intrinsic activity of TAFI [11,12]. Furthermore, the structure of bovine TAFIa in complex with tick carboxypeptidase inhibitor (TCI) was determined and found to exhibit a high degree of identify with the CPB-TCI structure [43-45]. Interestingly, the bovine TAFIa structure contains two undefined regions, both of which are part of exposed loops present in the Lβ2β3 and Lα2β4 regions and in a heparin affinity region [45]. The domains including Arg302 and Arg330, which are predicted to cause instability in human TAFI, were fully ordered in the bovine molecule.

These recent advances prompted us to perform a thorough biochemical characterization of the bovine protein, purified from bovine plasma. This biochemical characterization included analysis of stability, N-linked glycosylation, generation of TAFIa by removal of the pro-peptide by trypsin and thrombin/solulin, the antifibrinolytic effects of TAFIa, as well as analysis of the intrinsic activity of the full length protein and its potential to become crosslinked to fibrin by transglutaminases.

Results

Primary Structure of Bovine TAFI

The amino acid sequence of bovine TAFI was deduced from a cDNA library and published recently [45]. The sequence was 78.6% identical to that of the human protein. The bovine protein consisted of 401 amino acid residues, including a 92-amino acid residue pro-peptide that is released by cleavage at Arg92. All potential glycosylation sites were conserved and found glycosylated in both species, with exception of the fifth site (Asn219), which remained unglycosylated in bovine TAFI. The location of cysteine residues was identical in both species, with the exception of Cys69. This cysteine residue, which is located in the activation peptide, was absent from bovine TAFI. In human TAFI, Cys69 does not form a disulfide bridge and therefore, is unlikely to affect tertiary structure. All sites involved in catalysis as well as substrate and zinc binding were identical, suggesting that the two proteins have the same proteolytic properties.

Generation and Activity of TAFIa

SDS-PAGE of purified bovine TAFI produced a single sharp band at around 56 kDa, which is slightly lower than the position of human TAFI (Fig 1). TAFIa generated by either trypsin (Fig 1A) or thrombin/solulin complex (Fig 1B) migrated at the same position for both species, suggesting that differences in the migration of the full-length species can be attributed to differences in carbohydrates attached to the pro-peptide. It is obvious from the results of the SDS-PAGE that greater amounts of proteinases were required to generate human TAFIa (Fig 1). Since the thrombin/solulin complex is considered to be responsible for release of

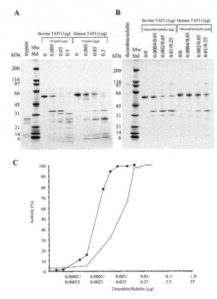

Figure 1. Generation of bovine and human TAFIa. Bovine and human TAFI (1 µg) were incubated with increasing amounts of trypsin (A) or thrombin/solulin complex (B) (all values in µg). Proteolysis products were then analyzed by SDS-PAGE and visualized by Coomassie Brilliant Blue staining. Additionally, TAFI (0.2 µg of bovine or human) was incubated with increasing amounts of thrombin/solulin complex (C). Increase in activity of bovine (filled circles) and human (open circles) TAFIa was monitored through HPLC based kinetic assay using Hip-Arg substrate as described in the method section. Note that compared to human TAFI, roughly 15 times less proteinase complex is required to generate 100% active bovine TAFIa.

the TAFI pro-peptide in vivo [14] and since trypsin seems to inactivate TAFIa more aggressively, only the complex was used to determine the optimal conditions for generation of bovine and human TAFIa. Increasing amounts of the thrombin/solulin were incubated with TAFI, and the enhanced activity was monitored by HPLC based activity assay. As shown in Fig 1C, the amount of proteinase complex required to achieve 100% TAFIa activity was much lower than that needed for the human TAFI. Furthermore, the kinetic constants for both species were determined (Table 1). The intrinsic activity of TAFI is similar between species, while the Vmax and Km for bovine TAFa is somewhat higher in comparison to the human protein.

Table 1. Summary of bovine and human TAFI and TAFIa kinetic values

Equation*	Human TAFI		Bovine TAFI		Human TAFIa		Bovine TAFIa	
	Vmax	Km	Vmax	Km	Vmax	Km	Vmax	Km
Hanes	40.17	3.41	31.67	4.63	106.00	3.14	330.67	7.86
Eadie-Hofstee	43.00	3.96	32.50	4.65	113.33	3.68	332.00	7.84
Eisenthal-Cornish-Bowden	40.83	3.91	33.33	4.69	112.67	3.74	329.33	8.24
Hyperbolic Regression	41.33 ± 7.93	3.70 ± 1.51	33.40 ± 6.33	4.85 ± 2.50	106.67 ± 8.34	2.83 ± 0.89	326.67 ± 39.67	5.28 ± 1.96
Average values	41.33	3.75	32.73	4.71	109.67	3.35	329.67	7.31
Kcat (min⁻¹)	243.12		192.53		4386.80		13186.80	
Kcat/Km (min⁻¹/mM)	64.83		40.88		1309.49		1803.94	

*The values for Km and Vmax were determined by the direct fit of the Michaelis-Menten equation employing four graphical methods. The data represent the enzyme-catalysed reaction for 0.33 µM TAFI and 0.025 µM TAFIa.

Identification of Proteolysis Products Generated Upon Proteinase Addition to TAFI

The thrombin/solulin complex produced similar proteolytic fragmentation in both the bovine and human TAFI. The generated products were identified by Edman degradation and are summarized in Table 2. SDS-PAGE of bovine TAFI and thrombin/solulin mixture fashioned a strong band not only at 56 kDa (corresponding to full length TAFI), but also at 36 kDa, which was confirmed to be TAFIa (Fig 2 and Table 2). In contrast to the human TAFI pro-peptide, the released bovine pro-peptide was clearly visible by Coomassie staining, with a mass of around 29 kDa (Fig 2). As expected, large amounts of thrombin/solulin complex truncated human TAFIa (36 kDa) at Arg[302], liberating the 11.0 kDa C-terminal peptide to produce a proteolytically inactivated form of TAFIa (24.7 kDa) (Fig 1B). However, no further proteolytic products were observed for the bovine protein in the higher end of the titration using the proteinase complex.

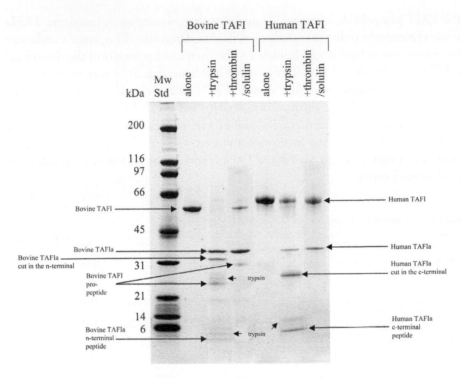

Figure 2. Identification of bovine TAFI products generated by proteolysis. SDS-PAGE of bovine or human TAFI (1 μg), which was cleaved using either 0.05 μg of trypsin, or thrombin/solulin complex in a ratio of 0.002 μg/0.05 μg. The products are indicated with arrows and were identified by Edman-degradation (see Table 2 for a summary of the bovine TAFI products).

Table 2. Summary of bovine TAFI activation products

Product**	Proteinase used in activation	SDS-PAGE mass (kDa)	Theoretical mass (kDa)	N-terminal sequence*
Bovine TAFI zymogen	N/A (full length TAFI)	56	46.4	FQRGVLSALP
Bovine TAFIa	trypsin solulin/thrombin	36	35.9	ASSSYYEQYH
Bovine TAFIa cut at the n-terminus	trypsin	32	29.2	AKNAMWID
Bovine TAFI activation peptide	trypsin solulin/thrombin	29	10.5	FQRGVLSALP
Bovine TAFI activation peptide	trypsin	25	NA	FQRGVLSALP
Bovine TAFIa N-terminal peptide	trypsin	<6	6.7	ASSSYYEQYH

*Determined by Edman Degradation.
**Upon cleavage of bovine TAFI by either trypsin or solulin/thrombin complex, fragments were separated by SDS-PAGE, transferred to PVDF membrane and subjected to Edman –degradation as described in the method section. The obtained n-terminal sequences of fragments and their molecular weight is listed.

When TAFI was cleaved by trypsin, which is more potent than thrombin/solulin complex, a somewhat dissimilar fragmentation pattern was generated (Fig 2). SDS-PAGE of the human protein yielded a typical pattern consisting of full length TAFI at 58 kDa, mature TAFIa at 36 kDa, C-terminal processed TAFIa (through cleavage at Arg330) at 28 kDa, and the released C-terminal peptide at 8 kDa (Fig 2). Trypsin cleavage occurred at the same site in both the human and bovine protein, creating a 36 kDa TAFIa through truncation at Arg92 (Fig 2). Interestingly, in contrast to human TAFIa, bovine TAFIa was initially proteolytically inactivated by cleavage at the N-terminus, rather than the C-terminus. This processing occurred right after Arg147, creating a 29.2 kDa fragment, detected around 32 kDa on SDS gel (Fig 2 and Table 2). The pro-peptide was detected at around 29 kDa (Fig 2). However, it is most likely immediately processed at the C-terminus, as a 25 kDa band was detected that also contained the TAFI N-terminal sequence (Fig 2). Trypsin and small trypsin fragments were also detected, along with the released 6.7 kDa TAFIa N-terminal peptide (Fig 2 and Table 2).

Bovine TAFIa Stability at 37°C

The thermal stability of TAFIa was investigated with HPLC based kinetic assays using Hip-Arg as a substrate and the thrombin/solulin complex as an activator (Fig 3). The half-life of bovine TAFIa was longer than that of human TAFIa. Human TAFIa activity decreased by 50% after 5 min, while this decrease in bovine TAFIa activity occurred at 10 min (Fig 3).

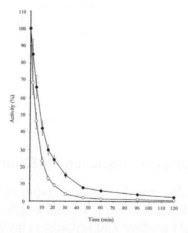

Figure 3. Bovine TAFIa is more stable at 37°C than human TAFIa. Bovine (filled circles) or human (open circles) TAFI (3 µg of each) were added thrombin/solulin complex (using optimal conditions as determined in the method section), placed at 37°C, and subjected to HPLC based kinetic assays using Hip-Arg substrate at the indicated intervals. Bovine TAFIa is more stable than human TAFIa, as seen by a half-life that is twice as long (i.e., 10 min vs. 5 min).

Intrinsic Activity of Bovine TAFI

The intrinsic activity of the bovine protein was determined by activity assays using the Hip-Arg substrate and compared to that of human TAFI, by measuring the released hippuric acid by HPLC. TAFI activity was undoubtedly detected using this method (Fig 4A). Moreover, hippuric acid release was blocked when TAFI was incubated with the substrate in the presence of 5 mM 1, 10-phenantroline, a chelating agent and known carboxypeptidase inhibitor (Fig 4B). This confirms that bovine TAFI has genuine intrinsic activity. This activity was also inhibited by 2.3 µM TCI, a potentially physiologically relevant inhibitor of TAFI (Fig 4C).

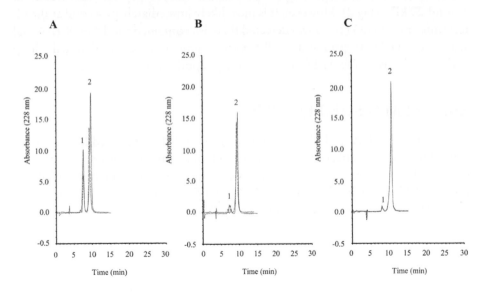

Figure 4. Bovine TAFI cleaves Hip-Arg substrate. The intrinsic activity of the bovine and human protein (1 µg) was investigated by incubating TAFI with Hip-Arg substrate in the absence (A) or presence of 5 mM 1, 10-phenantroline (B) or 1 µg TCI (C). The cleaved product, hippuric acid (1), was then separated from the internal standard (2) by RP-HPLC. Bovine TAFI, similar to human TAFI, produces considerable hippuric acid, and this carboxypeptidase activity is abolished by addition of either 1, 10-phenantroline or TCI.

Isoelectric Point Variation Between TAFI and TAFIa

Isoelectric focusing revealed that the isoelectric point of TAFI and TAFIa varied greatly in both species (data not shown). The full length bovine protein migrated in the lower end of the pH gradient and appeared as multiple bands between a pI of 6.0 and 6.5, likely due to glycosylation heterogeneity. This migration position was slightly higher than that of the human TAFI, which migrated to a pI of 5.1 to 6.0 and also appeared as multiple isoforms. Upon release of the heavily glycosylated activation peptide, both human and bovine TAFIa appeared as a single band at a much higher pI of around 8.5. Thus, the difference in migration between TAFI in the two species is due to differences in the carbohydrate modifications to their respective pro-peptides.

Bovine TAFI Attenuates Clot Lysis In Vitro

The effect of bovine TAFI on fibrinolysis was examined by conducting a fibrinolyses assay in a purified system (Fig 5). The clot was generated in microtiter wells by addition of thrombin to fibrinogen and the clot lysis initiated by further addition of plasminogen and tPA simultaneously. Purified bovine or human TAFI (1 µg) added to the wells, in the presence and absence of solulin, was able to delay clot lysis. Moreover, this effect was reversed by the carboxypeptidase inhibitors, PCI or TCI, confirming capability of bovine (and human) TAFI to effect clot lysis (data not shown).

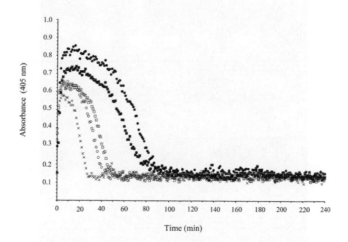

Figure 5. Bovine TAFIa attenuates clot lysis. The anti fibrinolytic function of bovine TAFIa was tested and compared to that of human TAFIa in a purified system. Clott formation was initiated by addition of thrombin to fibrinogen in the presence of CaCl2. Simultaneously, the dissolution of clot was generated by tPA and plasminogen addition. The change in turbidity was monitored at 405 nm (crosshairs). Upon addition of 1 µg of either bovine TAFI (filled circles) or human TAFI (filled squares), in the presence of Solulin, a delay in clot lysis was observed. A small anti fibrinolytic effect was observed upon addition of 1 µg of bovine TAFI (open circles) and human TAFI (open squares) in the absence of Solulin as well.

Bovine TAFI is a Substrate for Tissue Transglutaminase

To test whether bovine TAFI has the potential to become cross-linked to a fibrin meshwork in the same manner as the human protein, we monitored the incorporation of a fluorescent donor, dansylcadaverine, into the protein by tissue transglutaminase. Visualization of SDS-polyacrylamide gels with UV light revealed that both human TAFI and α-2AP (another known tissue transglutaminase substrate) incorporated dansylcadaverine (Fig 6). Importantly, dansylcadaverine was also successfully incorporated into bovine TAFI under these conditions, showing that bovine TAFI can serve as a transglutaminase substrate.

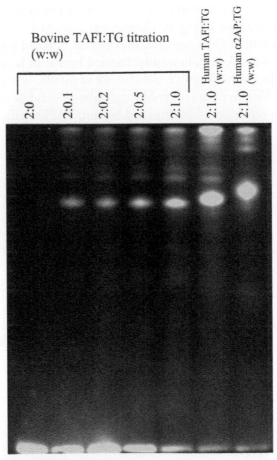

Figure 6. Bovine TAFI is a substrate for transglutaminases. Bovine TAFI (2 μg) was incubated with increasing amounts of tissue transglutaminase (TG), in the presence of the fluorescent donor, dansylcadaverine. The reaction products were separated by SDS-PAGE and visualized under UV light. Human TAFI (2 μg) and α2-antiplasmin (2 μg) served as a controls. Note the clear incorporation of dansylcadavarine into bovine TAFI, suggesting that it contains amine acceptor sites and functions as a substrate for transglutaminases.

Bovine TAFI Contains Four N-Linked Carbohydrate Structures, Which are Located Solely on the Activation Peptide

To characterize the glycans of bovine TAFI, we performed tryptic digests with and without subsequent PNGase F treatment, and separated the resulting fragments using RP-HPLC. Fractions containing glycopeptides were purified and analyzed using MALDI-TOF MS and subsequently verified based on their fragmentation using MALDI quadrupole (Q) TOF MS/MS (data not shown). Thirty-five N-glycans were observed from the four N-glycosylation sites present within the N-terminal activation region (i.e., N22, N51, N63, and N86) (Fig 7 and Table 3). Biantennary structures without core fucosylations were the sole structures identified by the glycoanalysis. Hence, the substantial microheterogeneity observed was limited to variations in the contents of the two types of sialic acids, N-glycolyl-neuraminic acid (Neu5Gc) and N-acetylneuraminic acid (Neu5Ac). Up to four sialic acid residues were observed on the N22 glycans, whereas the rest of the occupied sequons contained a maximum of three sialic acid residues (Fig 7). Peptide mass fingerprinting of tryptic bovine TAFI treated with and without PNGase F revealed that neither of the two potential sites located in the middle of the protein (N219 and N226) was occupied by N-glycans. In contrast, N-linked glycosylation has been detected in human TAFIa. The human protein also exhibits a much greater heterogeneity in the N-linked sugars [18]. This may help account for difficulties in determining the three-dimensional structure of human TAFI zymogen.

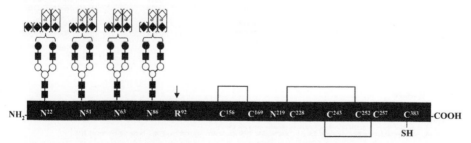

Figure 7. N-glycans of bovine TAFI. The site in the N-terminal sequence which is cleaved, releasing the pro-peptide (Arg^{92}) and known disulfide bonds of bovine TAFI are shown in the schematic, as well as the glycosylation sites (not drawn to scale). Square brackets indicate that the glycans were observed with and without the particular carbohydrate residue. (Filled squares, N-acetylglucosamine; Open circles, mannose; Filled circles, galactose; Filled diamonds, 5-N-acetylneuraminic acid; Open diamonds, 5-N-glycolylneuraminic acid).

Table 3. Structure of N-linked carbohydrates in bovine TAFI

Site	N-Glycans/Occupancy[**]	Observed glycopeptide mass (Da)	Theoretical glycopeptide mass (Da)	Glycan mass (Da)	Glycans[***]
N[22]	Yes/100%	3157.2	3157.4	1622.6	$Gal_2Man_3GlcNAc_4$
		3448.3	3448.5	1913.7	$Neu5Ac_1Gal_2Man_3GlcNAc_4$
		3464.3	3464.5	1929.7	$Neu5Gc_1Gal_2Man_3GlcNAc_4$
		3739.5	3739.6	2204.8	$Neu5Ac_2Gal_2Man_3GlcNAc_4$
		3755.5	3755.6	2220.8	$Neu5Gc_1Neu5Ac_1Gal_2Man_3GlcNAc_4$
		4030.6	4030.7	2495.9	$Neu5Ac_3Gal_2Man_3GlcNAc_4$
		4046.7	4046.7	2511.9	$Neu5Gc_1Neu5Ac_2Gal_2Man_3GlcNAc_4$
		4062.6	4062.7	2527.9	$Neu5Gc_2Neu5Ac_1Gal_2Man_3GlcNAc_4$
		4321.8	4321.8	2787.0	$Neu5Ac_4Gal_2Man_3GlcNAc_4$
		4337.8	4337.8	2803.0	$Neu5Gc_1Neu5Ac_3Gal_2Man_3GlcNAc_4$
		4353.8	4353.8	2819.0	$Neu5Gc_2Neu5Ac_2Gal_2Man_3GlcNAc_4$
N[51]	Yes/100%	3533.6	3533.5	1622.6	$Gal_2Man_3GlcNAc_4$
		3824.7	3824.6	1913.7	$Neu5Ac_1Gal_2Man_3GlcNAc_4$
		3840.7	3840.6	1929.7	$Neu5Gc_1Gal_2Man_3GlcNAc_4$
		4115.9	4115.7	2204.8	$Neu5Ac_2Gal_2Man_3GlcNAc_4$
		4131.9	4131.7	2220.8	$Neu5Gc_1Neu5Ac_1Gal_2Man_3GlcNAc_4$
		4407.1	4406.8	2495.9	$Neu5Ac_3Gal_2Man_3GlcNAc_4$
		4423.1	4422.8	2511.9	$Neu5Gc_1Neu5Ac_2Gal_2Man_3GlcNAc_4$
		4439.1	4438.8	2527.9	$Neu5Gc_2Neu5Ac_1Gal_2Man_3GlcNAc_4$
N[63]	Yes/100%	2389.9	2390.0	1622.6	$Gal_2Man_3GlcNAc_4$
		2681.0	2681.1	1913.7	$Neu5Ac_1Gal_2Man_3GlcNAc_4$
		2697.0	2697.1	1929.7	$Neu5Gc_1Gal_2Man_3GlcNAc_4$
		2972.1	2972.2	2204.8	$Neu5Ac_2Gal_2Man_3GlcNAc_4$
		2988.0	2988.2	2220.8	$Neu5Gc_1Neu5Ac_1Gal_2Man_3GlcNAc_4$
		3263.3	3263.3	2495.9	$Neu5Ac_3Gal_2Man_3GlcNAc_4$
		3279.2	3279.3	2511.9	$Neu5Gc_1Neu5Ac_2Gal_2Man_3GlcNAc_4$
		3295.3	3295.3	2527.9	$Neu5Gc_2Neu5Ac_1Gal_2Man_3GlcNAc_4$
N[84]	Yes/100%	2851.0	2851.2*	1622.6	$Gal_2Man_3GlcNAc_4$
		3142.1	3142.3*	1913.7	$Neu5Ac_1Gal_2Man_3GlcNAc_4$
		3158.1	3158.3*	1929.7	$Neu5Gc_1Gal_2Man_3GlcNAc_4$
		3433.2	3433.4*	2204.8	$Neu5Ac_2Gal_2Man_3GlcNAc_4$
		3449.2	3449.4*	2220.8	$Neu5Gc_1Neu5Ac_1Gal_2Man_3GlcNAc_4$
		3724.3	3724.5*	2495.9	$Neu5Ac_3Gal_2Man_3GlcNAc_4$
		3740.4	3740.5*	2511.9	$Neu5Gc_1Neu5Ac_2Gal_2Man_3GlcNAc_4$
		3756.4	3756.5*	2527.9	$Neu5Gc_2Neu5Ac_1Gal_2Man_3GlcNAc_4$
N[219]	unglycosylated				
N[226]	unglycosylated				

*Theoretical glycopeptide mass values are based on the transformation of the N-terminal Gln[82] to pyroglutamic acid.
**Percent occupancy of N-glycans was determined based on MS experiments in which peptide mass fingerprints of tryptic bovine TAFI treated ± PNGase F were compared to establish the presence of (non/de)-glycosylated peptides (data not shown).
***GlcNAc, N-acetylglucosamine; Man, mannose; Gal, galactose; Neu5Ac, 5-N-acetylneuraminic acid; Neu5Gc, 5-N-glycolylneuraminic acid.

Discussion

The structure of TAFI and TAFIa/TCI complex has recently been solved using authentic bovine TAFI [42,45]. Here, we present a full biochemical characterization of the bovine protein purified from bovine plasma. The amino acid sequence idenity between the two spieces is 78.6% and all the important sites, such as the catalytic domain, substrate-binding domain, and zinc-binding domain, are fully conserved [45]. Only four of the potential N-linked carbohydrate sites are occupied, all located on the activation peptide. The fifth site, Asn^{219}, which is partially glycosylated in the human protein [18], remained unglycosylated in the bovine protein (Fig 7). Accordingly, this site was found to be buried within the protein structure and has previously been suggested to be unglycosylated [42]. Recently, the biochemical importance of human TAFI glycosylation has been studied using TAFI mutants [46]. Interestingly, in some mutants, the absence of carbohydrates increases the activity of full length TAFI, but decreases TAFIa activity. The increase in intrinsic activity is most apparent in the mutants TAFI-N22Q and TAFI-N22Q-N51Q-N63Q. These observations corroborate the finding that, in human TAFI, access to the active site exists [11] and this access site potentially expands upon carbohydrate removal, possibly imparting a catalytic function to sugars [46].

Interestingly, the pronounced microheterogeneity of the TAFI glycans was exclusively generated by the variation in the number and type of sialic acid residues located in the termini of the biantennary complex glycans. Neu5Ac and Neu5Gc were found in the TAFI glycans and both are known to be abundant sialic acids in bovine glycoconjugates. In contrast, humans cannot synthesize Neu5Gc, highlighting a difference between the authentic human and authentic bovine TAFI structure.

Purified bovine TAFI successfully attenuated fibrinolysis of tPA-induced clots in a purified system. Also similar to human TAFI, the bovine protein displays considerable stable intrinsic activity, which can be abolished by the same inhibitors used to inhibit TAFIa. Furthermore, it is most likely crosslinked to the fibrin meshwork during the early stages of fibrinolysis, as the protein seems to act as a substrate for transglutaminases. Bovine TAFI contains potential amine acceptor sites, as evidenced by the successful incorporation of dansylcadavarine into the protein by tissue transglutaminase.

Bovine TAFI, like the human protein, can also be cleaved through proteolysis at Arg92, generating the mature form, TAFIa. In contrast to the human TAFI, bovine TAFI is processed into not only the 36 kDa active enzyme, but also a 29 kDa TAFIa fragment following incubation with trypsin. This N-terminally processed TAFIa is missing a 7 kDa N-terminal peptide and is formed through proteolysis of the Arg147-Ala148 bond. This cleavage takes place prior to the usual inactivation that occurs at the C-terminus. Human TAFI contains either a Thr or Ala at position 147, depending on the variant. Therefore, an identical N-terminal truncation at this position is not possible [47]. Similar fragmentation has been observed following activation of rat TAFI by plasmin [35]. This may also explain the disordered Lβ2β3 segment observed in the three-dimensional structure of the TAFIa/TCI complex [45].

In human TAFI, substituting His333 with Tyr or Gln increases the half-life of TAFIa for up to 1.5 h, while preserving all characteristics of wild type TAFI [48]. Site-directed mutagenesis of Arg302, Arg320, and Arg330 produces a molecule much less stable than the wild type protein, suggesting that this instability is concentrated in the 302–330 region [21,49]. The naturally occurring mutation of Thr325 to Ile325 has been shown to make human TAFI twice as stable [21,50,51]. Position 325 of the bovine protein is occupied by Ile, which might account for the longer half-life of bovine TAFIa (10 min) compared to the human TAFIa (5 min). Similarly, mutation of Thr329 to Ile329 increases not only the half-life of the cleaved human protein, but also its fibrinolytic effect [21]. Again, this position is occupied by Ile in bovine TAFI.

Substitution of human TAFI residues with corresponding residues of CPB, such as TAFIa-Ile182Arg-Ile183Glu, does not significantly increase stability. On the contrary, it reduces antifibrinolytic potential. Nevertheless, lower amounts of thrombin-thrombomodulin complex are required in order to generate TAFIa from this mutant [52]. This can explain, at least partly, why lower amounts of proteinases are required to generate bovine TAFIa, in which Lys182 and Glu183 naturally occur in sequence. Indeed, 15 times less solulin/thrombin complex is required to generate bovine TAFIa with activity similar to that of human TAFIa.

In summary, we deduce that human TAFI and bovine TAFI have similar properties. The overall secondary structure is conserved, generation of TAFIa can be achieved in similar manner, and bovine TAFIa produces a measurable effect on fibrinolysis. Thus, the available three-dimensional structure of bovine TAFI is a reliable model for investigation of human TAFI, including its in vivo function and the in vivo effects of its inhibition.

Conclusion

The bovine and human TAFI activation occurs at equivalent sites and both TA-FIa and TAFI exhibit caroboxypeptidase activity. Additionally, TAFI from both species was found to be substrate for transglutaminases. Minor differences in the enzymatic stability of bovine and human TAFIa was observed as well as differences in the level of glycosylation, isoelectric point and proteolytic by-products in trypsin activation. However, overall the findings suggested that the the two orthologous proteins are similar and that conclusions reached using the bovine TAFI can safely be extrapolated to the human protein.

Methods

Materials

Bovine trypsin, 1, 10-phenantroline, phenylmethylsulfonyl fluoride (PMSF), polyethylene glycol 8000 (PEG), and the chromogenic carboxypeptidase substrate, hippuryl-Arg (Hip-Arg), were obtained from Sigma. Ortho-methylhippuric acid and Pefablock SC were from Aldrich. ECH-Lysine Sepharose was from Amersham Biosciences, GE Healthcare (Uppslala, Sweden). Dansylcadaverine was from Molecular Probes (Eugene, OR).

Proteins

Human TAFI and human α_2-antiplasmin (α-$_2$AP) were purified from normal human plasma (Statens Serum, Institute, Copenhagen, Denmark using plasminogen-depleted plasma and plasminogen-Sepharose affinity chromatography as described previously [4,18]. Guinea pig liver (tissue) transglutaminase (EC 2.3.2.13), human fibrinogen and human thrombin (EC 3.4.21.5) were purchased from Sigma. Recombinanat tPA (EC 3.4.21.68) was purchased from ProSpec-Tany TechnoGene LTD., Rehovot, Israel. Recombinant soluble thrombomodulin (solulin) was a generous gift of Dr. Achim Schuettler (PAION GmbH, Aachen, Germany) and Factor XIIIa from Sanofi-Aventis. Potato carboxypeptidase inhibitor (PCI) and TCI were kind gifts from Prof. Francesc. X. Aviles, Dr. Joan Lopez Arolas, and Dr. Laura Sanglas. TAFI-antiserum was raised commercially (Pel-Freez, Rogers, AR). Human plasminogen was purified by affinity chromatography using ECH-Lysine Sepharose as described previously [53].

Purification of Bovine TAFI

Bovine TAFI was purified essentially as already described [45]. In short, bovine blood (10 L) was collected at the local slaughterhouse and supplemented with 5 mM EDTA to prevent coagulation. The plasma was separated from erythrocytes by centrifugation at 600 × g for 15 min at 22°C. Plasma was incubated with 6% (w/v) PEG, and after 1 h, the precipitated proteins were removed by centrifugation at 10,000 × g for 40 min at 4°C. Plasminogen was removed from the supernatant by affinity chromatography using 1 L of ECH-Lysine Sepharose equilibrated in binding buffer (50 mM NaH_2PO_4, pH 7.5 and 100 mM NaCl). Plasminogen-depleted plasma was applied to a 500-ml plasminogen Sepharose column equilibrated in binding buffer, and bovine TAFI was eluted using 50 mM γ-amino-caproic acid. After buffer exchange into 20 mM Tris-Cl (pH 7.5), bovine

TAFI was separated from other contaminants by ion-exchange chromatography on a 5-ml HiTrapQ column connected to an AKTA Prime system (Amersham Biosciences, GE Healthcare). The column was eluted, at a flow rate of 1 ml/min, using a 0.5%/min linear gradient of Buffer A (20 mM Tris-Cl, pH 7.5) and Buffer B (20 mM Tris-Cl, pH 7.5 containing 1 M NaCl).

Polyacrylamide Gel Electrophoresis

Proteins were separated by SDS-PAGE in 5 – 15% polyacrylamide gels [54]. Samples were boiled for 5 min in the presence of 30 mM dithiothreitol (DTT) and 1% SDS prior to electrophoresis.

Generation of Human and Bovine TAFIa

Human and bovine TAFI (1 µg) were incubated with increasing amounts of the thrombin/solulin complex, (0 µg/0 µg to 0.01 µg/0.25 µg) for 30 min at 22°C in 20 mM Tris-HCl and 100 mM NaCl, pH 7.5. For trypsin induced proteolysis, 1 µg TAFI (human and bovine) was incubated with increasing amounts of trypsin (0–0.5 µg) for 20 min at 37°C in 20 mM Tris-HCl and 100 mM NaCl, pH 7.5. All reactions were terminated by addition of Pefablock or PMSF to a final concentration of 5 mM. Optimal conditions to generate TAFIa with peak activity were determined through kinetic assays (using 0.2 µg of TAFI) and SDS-PAGE (using 1.0 µg TAFI) with the physiologically relevant thrombin/solulin complex as an activator only. To activate 1 µg of human TAFI, the optimal trombin/solulin complex ratio (w/w) was 0.06 µg/1.5 µg. Generation of bovine TAFIa was optimal using thrombin/solulin complex ratio of 0.004 µg/0.1 µg to 1 µg TAFI.

NH$_2$-Terminal Amino Acid Sequencing

Proteolytic fragments of bovine TAFI generated by trypsin or solulin/thrombin complex were separated by SDS/PAGE. The stacking gel was allowed to polymerize one day prior to electrophoresis, and samples were heated for 3 min at only 80°C prior to separation. After electrophoresis, proteins were transferred to a polyvinylidene difluoride membrane (Immobilon-P, Millipore) in 10 mM CAPS and 10% (v/v) methanol (pH 11) [55]. Alternatively, the TAFI-trypsin or TAFI-solulin/thrombin mixture was applied to an activated ProSorb sample preparation cartridge (Applied Biosystems), according to the manufacturer's instructions. Samples were analyzed by automated Edman degradation using an Applied Biosystems PROCISE™ 494 HT sequencer with on-line HPLC (Applied Biosystems Model 120A) for phenylthiohydantoin analysis.

Isoelectric Focusing

Isoelectric focusing of bovine TAFI was performed essentially as described previously [18]. Briefly, 10 μg of salt-free protein was focused under native conditions in a Ready IEF gel using the MiniProtean III Cell (Biorad) according to the manufacturer's instructions. Bands were focused in a pH gradient of 3 – 10 using 20 mM lysine and 20 mM arginine as a cathode buffer and 7% phosphoric acid as anode buffer (all in H2O). Running conditions consisted of 100 V for 60 min, 250 V for 60 min, and 500 V for 30 min. Bands were visualized using IEF staining solution (27% isopropyl alcohol, 10% acetic acid, 0.04% Coomassie Blue R250, and 0.05% Crocein Scarlet in H2O).

HPLC Based Kinetic Activity Assay Using Hip-Arg Substrate

The activity of both full length and mature TAFI was determined essentially as described previously [56]. A 10-μl sample containing 1 μg of bovine TAFI or 0.2 μg TAFIa was incubated with 40 μl of 30 mM Hip-Arg for 40 min. Some samples were incubated for 15 min with 5 mM phenanthroline or 1 μg TCI prior to substrate addition. The reactions were stopped by addition of 50 μl 1 M HCl. Ten microliters of 15 mM ortho-methylhippuric acid was included in the reaction mixture as an internal standard. The reaction products, as well as the internal standard, were extracted using 300 μl ethyl acetate. One-hundred microliters of the extracted sample were lyophilized, solubilized in 100 μl mobile phase buffer [10 mM KH2PO4, pH 3.4 containing 15% acetonitrile (ACN)], and separated on a reverse phase (RP) HPLC column (PTH C18, 5 μm, 220 × 2.1 mm, Applied Biosystems) using the ÄKTA Ettan system (Amersham Biosciences, GE Healthcare).

Determination of TAFI Kinetic Constants

Human and bovine TAFI kinetic properties were essentially determined as described previously, with small modifications [46]. Briefly, 1 μg of the zymogen and 0.1 μg of TAFIa, generated by the thrombin/solulin complex, for both human and bovine protein, were incubated with increasing concentration of the Hip-Arg substrate (0–30 mM), in duplicates, for 60 min at 37°C in a final volume of 60 μl. The reaction was terminated by addition of 20 μl 1 M HCl, neutralized by addition of 20 μl of 1 M NaOH and buffered with 25 μl of 1 M NaH$_2$PO$_4$, pH 7.4. Upon addition of 60 μl 6% cyanuric chloride dissolved in 1,4-dioxane, the samples were vortexed vigorously and centrifuged at 16000 × g for 5 minutes. The supernatant was subsequently transferred to 96-well microtiter plate and the absorbance was measured at 405 nm in a FLUOStar Omega plate reader (BMG

Labtech) using the endpoint mode. The kinetic constants were determined using 4 different graphical methods.

Thermal Stability of TAFIa Enzymatic Activity

Bovine and human TAFI (3 µg) were mixed with solulin/thrombin complex using the optimal conditions for generation of TAFIa for each species. At the time of reaction termination with pefablock (5 mM final concentration), the reaction mixture was placed at 37°C. At various intervals over 120 min, 0.2 µg of TAFI protein was removed and incubated with Hip-Arg substrate. Kinetic measurements were then performed using HPLC method described above.

In Vitro Clot Lysis Assays

Clot lysis assays were performed essentially as described previously [57] using 96-well microtiter plates, with some modifications. Twenty µl of fibrinogen (20 µl/µg), 1 µl of plasminogen (0.5 µg/µl) and 12.5 µl of factor XIIIa (0.8 µg/µl) were mixed in a final volume of 100 µl in 20 mM Hepes and 150 mM NaCl, 5 mM $CaCl_2$, pH 7.4 (reaction buffer) in a set of wells. In a proximate set of wells, 1 µl of tPA (0.002 µg/µl) and 2 µl of thrombin (20 U/ml) were combined in a final volume of 50 µl using the reaction buffer. Clotting was initiated by addition of 50 µl of the fibrinogen/plasminogen/factor XIIIa mixture to wells containing tPA and thrombin. In some wells, 10 µl of solulin (0.1 µg/µl) was added to the tPA/thrombin mixture prior to clot initiation. Purified human or bovine TAFI (1 µg), was added to certain wells containing tPA, thrombin and (+/-) solulin, moments prior to the start of the clotting generation. Some wells contained additionally 1.27 µM TCI or 4.65 µM PCI. The turbidity of the clot was measured continuously at 405 nm in a plate reader (FLUOstar Omega, BMG LABTECH GmbH) at 37°C. The lysis time was defined as the time required for a 50% reduction in optical density.

Incorporation of Dansylcadaverine Using Tissue Transglutaminase

Human TAFI, bovine TAFI, or α-2AP (2 µg of each) were incubated with varying amounts of tissue transglutaminase (0 – 2 µg) for 3 h at 37°C in 20 mM Tris-Cl and 100 mM NaCl (pH 7.5) containing 10 mM Ca^{2+}, 0.5 mM DTT, and 0.5 mM dansylcadaverine. The reaction was stopped by addition of 10 mM EDTA, and samples were analyzed by reducing SDS-PAGE. The gel was visualized under UV light.

Amino Acid Sequence Analysis

To determine the accurate concentration of TAFI used in this study, we performed each analysis in triplicate. For each analysis, approximately 2 µg of purified bovine or human TAFI was dried in 500 µl polypropylene vials. The lids were punctured, and the vials were placed in a 25-ml glass vial equipped with a MinInert valve (Pierce Biotechnology, Rockford, IL, USA). Two-hundred microliters of 6 N HCl containing 0.1% phenol was placed in the bottom of the glass and blown with argon before a vacuum was applied. The samples were incubated at 110°C for 18 h. They were subsequently redissolved in 50 µl 0.20 M sodium citrate loading buffer, pH 2.20 (Biochrom, Cambridge, UK), transferred to microvials, and loaded on a BioChrom 30 amino acid analyzer (Biochrom). Data analysis was performed using software developed in house.

Proteolytic Digestion of Bovine TAFI and Purification of Glycosylated Peptides

Modified trypsin (2 µg, Promega, Madison, WI) was added to approximately 40 µg of purified bovine TAFI in 20 mM Tris and 200 mM NaCl (pH 7.5) and then incubated overnight at 37°C. The resulting peptide mixture was split into two samples. N-glycosidase F (1 U, Roche, Mannheim, Germany) was added to one of the samples and incubated overnight at 37°C. The other sample was stored at -18°C. The two samples were applied separately to a reversed phase HPLC column (Jupiter C18 250 mm × 2 mm, 5 µm, 300 Å, Phenomenex, Torrance, CA) connected to an ÄKTA Basic instrument (Amersham Pharmacia Biotech, GE, Uppsala, Sweden). The sample was applied in buffer A [0.06% trifluoroacetic acid (TFA) in water] and eluted using the following three-step gradient in buffer B (0.05% TFA and 90% ACN in water): 5 to 40% in 30 min, 40 to 60% in 5 min, and 60 to 90% in 3 min. Differences in the corresponding chromatograms revealed the fractions potentially containing glycopeptides. These fractions were dried and redissolved in 5% formic acid for further analysis.

Characterization of Glycosylated Peptides by Matrix-Assisted Laser Desorption/Ionization Time-Of-Flight Mass Spectrometry (MALDI-TOF MS)

The fractions containing glycopeptides were concentrated and desalted using hydrophobic microcolumns packed with Poros R2 (20 µm, Applied Biosystems, Framingham, MA) in GelLoader pipette tips (Eppendorf, Hamburg, Germany) as described elsewhere [58]. The samples were eluted directly onto the MS target with 0.5 µl 2,5-dihydroxybenzoic acid (20 g/L) in 70% ACN and 0.1% TFA. Alternatively, fractions were not desalted and analyzed by mixing 0.5 µl sample and 0.5 µl matrix directly on the target. All samples were analyzed in positive polarity mode by MALDI MS using a Bruker Ultraflex (Bruker Daltonics, Bremen, Germany) with TOF-TOF technology or a MALDI Q-TOF Ultima (Waters, Micromass, Manchester, UK). The spectra were internally calibrated, or external calibration was performed by placing a tryptic lactoglobulin digest near the actual target spot.

Abbreviations

[1]The abbreviations used are: α-2AP: α2-antiplasmin; ACN, acetonitrile; Gal: galactose; Neu5Gc: 5-N-glycolylneuraminic acid; Neu5Ac: 5-N-acetylneuraminic acid; GlcNAc: N-acetylglucosamine; Hip-Arg: hippuryl-arginine; PMSF: Phenylmethanesulfonyl fluoride; Lys: lysine; Arg: arginine; Man: mannose; MALDI-TOF MS: matrix assisted laser desorption ionization time-of-flight mass spectrometry; PAGE: polyacrylamide gel electrophoresis; PCI: potato carboxypeptidase inhibitor; pI: isolectric point; PVDF: polyvinylidene difluoride; RP-HPLC: reverse phase high performance liquid chromatography; TAFI: zymogen of thrombin activatable fibrinolysis inhibitor; TAFIa: activated form of thrombin activatable fibrinolysis inhibitor; TFA: trifluoroacetic acid; CPB: carboxypeptidase B; tPA: tissue plasminogen activator; TCI: tick carboxypeptidase inhibitor.

Authors' Contributions

ZV performed the majority of the experimental work and wrote the manuscript. MTA and PH performed the carbohydrate analysis. Additionally they provided valuble suggestions and feedback prior to submission of the manuscript. KS and TCH assisted during the purification of the protein. TK cloned the bovine TAFI cDNA and provided the sequence. JJE supervised the experimental work, revised and finalized the manuscript. All authors read and approved the final manuscript.

Acknowledgements

We gratefully acknowledge the generous gift of TCI and PCI from Drs. F. X. Aviles, J. L. Arolas, and L. C. Sanglas as well as recombinant soluble thrombomodulin (solulin) from Dr. Achim Schuettler and Clemens Gillen (PAION

GmbH, Aachen Germany). The work was supported by grants from the Danish Natural Science Research Council (J.J.E.).

References

1. Hendriks D, Scharpe S, van Sande M, Lommaert MP: A labile enzyme in fresh human serum interferes with the assay of carboxypeptidase N. Clin Chem 1989, 35(1):177.

2. Hendriks D, Scharpe S, van Sande M, Lommaert MP: Characterisation of a carboxypeptidase in human serum distinct from carboxypeptidase N. J Clin Chem Clin Biochem 1989, 27(5):277–285.

3. Hendriks D, Wang W, Scharpe S, Lommaert MP, van Sande M: Purification and characterization of a new arginine carboxypeptidase in human serum. Biochim Biophys Acta 1990, 1034(1):86–92.

4. Eaton DL, Malloy BE, Tsai SP, Henzel W, Drayna D: Isolation, molecular cloning, and partial characterization of a novel carboxypeptidase B from human plasma. J Biol Chem 1991, 266(32):21833–21838.

5. Redlitz A, Tan AK, Eaton DL, Plow EF: Plasma carboxypeptidases as regulators of the plasminogen system. J Clin Invest 1995, 96(5):2534–2538.

6. Wang W, Hendriks DF, Scharpe SS: Carboxypeptidase U, a plasma carboxypeptidase with high affinity for plasminogen. J Biol Chem 1994, 269(22):15937–15944.

7. Bajzar L, Manuel R, Nesheim ME: Purification and characterization of TAFI, a thrombin-activable fibrinolysis inhibitor. J Biol Chem 1995, 270(24):14477–14484.

8. Sakharov DV, Plow EF, Rijken DC: On the mechanism of the antifibrinolytic activity of plasma carboxypeptidase B. J Biol Chem 1997, 272(22):14477–14482.

9. Mosnier LO, Meijers JC, Bouma BN: Regulation of fibrinolysis in plasma by TAFI and protein C is dependent on the concentration of thrombomodulin. Thromb Haemost 2001, 85(1):5–11.

10. Bouma BN, Meijers JC: Thrombin-activatable fibrinolysis inhibitor (TAFI, plasma procarboxypeptidase B, procarboxypeptidase R, procarboxypeptidase U). J Thromb Haemost 2003, 1(7):1566–1574.

11. Valnickova Z, Thogersen IB, Potempa J, Enghild JJ: Thrombin-activable fibrinolysis inhibitor (TAFI) zymogen is an active carboxypeptidase. J Biol Chem 2007, 282(5):3066–3076.

12. Willemse JL, Polla M, Hendriks DF: The intrinsic enzymatic activity of plasma procarboxypeptidase U (TAFI) can interfere with plasma carboxypeptidase N assays. Anal Biochem 2006, 356(1):157–159.

13. Valnickova Z, Enghild JJ: Human procarboxypeptidase U, or thrombin-activable fibrinolysis inhibitor, is a substrate for transglutaminases. Evidence for transglutaminase-catalyzed cross-linking to fibrin. J Biol Chem 1998, 273(42):27220–27224.

14. Bajzar L, Morser J, Nesheim M: TAFI, or plasma procarboxypeptidase B, couples the coagulation and fibrinolytic cascades through the thrombin-thrombomodulin complex. J Biol Chem 1996, 271(28):16603–16608.

15. Tan AK, Eaton DL: Activation and characterization of procarboxypeptidase B from human plasma. Biochemistry 1995, 34(17):5811–5816.

16. Mao SS, Cooper CM, Wood T, Shafer JA, Gardell SJ: Characterization of plasmin-mediated activation of plasma procarboxypeptidase B. Modulation by glycosaminoglycans. J Biol Chem 1999, 274(49):35046–35052.

17. Marx PF, Dawson PE, Bouma BN, Meijers JC: Plasmin-mediated activation and inactivation of thrombin-activatable fibrinolysis inhibitor. Biochemistry 2002, 41(21):6688–6696.

18. Valnickova Z, Christensen T, Skottrup P, Thogersen IB, Hojrup P, Enghild JJ: Post-translational modifications of human thrombin-activatable fibrinolysis inhibitor (TAFI): evidence for a large shift in the isoelectric point and reduced solubility upon activation. Biochemistry 2006, 45(5):1525–1535.

19. Valnickova Z, Thogersen IB, Christensen S, Chu CT, Pizzo SV, Enghild JJ: Activated human plasma carboxypeptidase B is retained in the blood by binding to alpha2-macroglobulin and pregnancy zone protein. J Biol Chem 1996, 271(22):12937–12943.

20. Wang W, Boffa MB, Bajzar L, Walker JB, Nesheim ME: A study of the mechanism of inhibition of fibrinolysis by activated thrombin-activable fibrinolysis inhibitor. J Biol Chem 1998, 273(42):27176–27181.

21. Boffa MB, Bell R, Stevens WK, Nesheim ME: Roles of thermal instability and proteolytic cleavage in regulation of activated thrombin-activable fibrinolysis inhibitor. J Biol Chem 2000, 275(17):12868–12878.

22. Marx PF, Hackeng TM, Dawson PE, Griffin JH, Meijers JC, Bouma BN: Inactivation of active thrombin-activatable fibrinolysis inhibitor takes place by a process that involves conformational instability rather than proteolytic cleavage. J Biol Chem 2000, 275(17):12410–12415.

23. Campbell WD, Lazoura E, Okada N, Okada H: Inactivation of C3a and C5a octapeptides by carboxypeptidase R and carboxypeptidase N. Microbiol Immunol 2002, 46(2):131–134.

24. Myles T, Nishimura T, Yun TH, Nagashima M, Morser J, Patterson AJ, Pearl RG, Leung LL: Thrombin activatable fibrinolysis inhibitor, a potential regulator of vascular inflammation. J Biol Chem 2003, 278(51):51059–51067.

25. te Velde EA, Wagenaar GT, Reijerkerk A, Roose-Girma M, Borel Rinkes IH, Voest EE, Bouma BN, Gebbink MF, Meijers JC: Impaired healing of cutaneous wounds and colonic anastomoses in mice lacking thrombin-activatable fibrinolysis inhibitor. J Thromb Haemost 2003, 1(10):2087–2096.

26. Hori Y, Gabazza EC, Yano Y, Katsuki A, Suzuki K, Adachi Y, Sumida Y: Insulin resistance is associated with increased circulating level of thrombin-activatable fibrinolysis inhibitor in type 2 diabetic patients. J Clin Endocrinol Metab 2002, 87(2):660–665.

27. Yano Y, Kitagawa N, Gabazza EC, Morioka K, Urakawa H, Tanaka T, Katsuki A, Araki-Sasaki R, Hori Y, Nakatani K, et al.: Increased plasma thrombin-activatable fibrinolysis inhibitor levels in normotensive type 2 diabetic patients with microalbuminuria. J Clin Endocrinol Metab 2003, 88(2):736–741.

28. Hataji O, Taguchi O, Gabazza EC, Yuda H, D'Alessandro-Gabazza CN, Fujimoto H, Nishii Y, Hayashi T, Suzuki K, Adachi Y: Increased circulating levels of thrombin-activatable fibrinolysis inhibitor in lung cancer patients. Am J Hematol 2004, 76(3):214–219.

29. Van Thiel DH, George M, Fareed J: Low levels of thrombin activatable fibrinolysis inhibitor (TAFI) in patients with chronic liver disease. Thromb Haemost 2001, 85(4):667–670.

30. Kremer Hovinga JA, Franco RF, Zago MA, Ten Cate H, Westendorp RG, Reitsma PH: A functional single nucleotide polymorphism in the thrombin-activatable fibrinolysis inhibitor (TAFI) gene associates with outcome of meningococcal disease. J Thromb Haemost 2004, 2(1): 54–57.

31. Klement P, Liao P, Bajzar L: A novel approach to arterial thrombolysis. Blood 1999, 94(8):2735–2743.

32. Kato T, Akatsu H, Sato T, Matsuo S, Yamamoto T, Campbell W, Hotta N, Okada N, Okada H: Molecular cloning and partial characterization of rat procarboxypeptidase R and carboxypeptidase N. Microbiol Immunol 2000, 44(8):719–728.

33. Marx PF, Wagenaar GT, Reijerkerk A, Tiekstra MJ, van Rossum AG, Gebbink MF, Meijers JC: Characterization of mouse thrombin-activatable fibrinolysis inhibitor. Thromb Haemost 2000, 83(2):297–303.

34. Bjorkman JA, Abrahamsson TI, Nerme VK, Mattsson CJ: Inhibition of carboxypeptidase U (TAFIa) activity improves rt-PA induced thrombolysis in a dog model of coronary artery thrombosis. Thromb Res 2005, 116(6):519–524.

35. Hillmayer K, Macovei A, Pauwels D, Compernolle G, Declerck PJ, Gils A: Characterization of rat thrombin-activatable fibrinolysis inhibitor (TAFI)–a comparative study assessing the biological equivalence of rat, murine and human TAFI. J Thromb Haemost 2006, 4(11):2470–2477.

36. Wang X, Smith PL, Hsu MY, Ogletree ML, Schumacher WA: Murine model of ferric chloride-induced vena cava thrombosis: evidence for effect of potato carboxypeptidase inhibitor. J Thromb Haemost 2006, 4(2):403–410.

37. Nagashima M, Yin ZF, Broze GJ Jr, Morser J: Thrombin-activatable fibrinolysis inhibitor (TAFI) deficient mice. Front Biosci 2002, 7:d556–568.

38. Nagashima M, Yin ZF, Zhao L, White K, Zhu Y, Lasky N, Halks-Miller M, Broze GJ Jr, Fay WP, Morser J: Thrombin-activatable fibrinolysis inhibitor (TAFI) deficiency is compatible with murine life. J Clin Invest 2002, 109(1):101–110.

39. Barbosa Pereira PJ, Segura-Martin S, Oliva B, Ferrer-Orta C, Aviles FX, Coll M, Gomis-Ruth FX, Vendrell J: Human procarboxypeptidase B: three-dimensional structure and implications for thrombin-activatable fibrinolysis inhibitor (TAFI). J Mol Biol 2002, 321(3):537–547.

40. Reeck GR, Neurath H: Isolation and characterization of pancreatic procarboxypeptidase B and carboxypeptidase B of the African lungfish. Biochemistry 1972, 11(21):3947–3955.

41. Marx PF, Brondijk TH, Plug T, Romijn RA, Hemrika W, Meijers JC, Huizinga EG: Crystal structures of TAFI elucidate the inactivation mechanism of activated TAFI: a novel mechanism for enzyme autoregulation. Blood 2008, 112(7):2803–2809.

42. Anand K, Pallares I, Valnickova Z, Christensen T, Vendrell J, Wendt KU, Schreuder HA, Enghild JJ, Aviles FX: The Crystal Structure of Thrombin-activable Fibrinolysis Inhibitor (TAFI) Provides the Structural Basis for Its Intrinsic Activity and the Short Half-life of TAFIa. J Biol Chem 2008, 283(43):29416–29423.

43. Arolas JL, Lorenzo J, Rovira A, Castella J, Aviles FX, Sommerhoff CP: A carboxypeptidase inhibitor from the tick Rhipicephalus bursa: isolation, cDNA cloning, recombinant expression, and characterization. J Biol Chem 2005, 280(5):3441–3448.

44. Arolas JL, Popowicz GM, Lorenzo J, Sommerhoff CP, Huber R, Aviles FX, Holak TA: The three-dimensional structures of tick carboxypeptidase inhibitor in complex with A/B carboxypeptidases reveal a novel double-headed binding mode. J Mol Biol 2005, 350(3):489–498.

45. Sanglas L, Valnickova Z, Arolas JL, Pallares I, Guevara T, Sola M, Kristensen T, Enghild JJ, Aviles FX, Gomis-Ruth FX: Structure of activated thrombin-activatable fibrinolysis inhibitor, a molecular link between coagulation and fibrinolysis. Mol Cell 2008, 31(4):598–606.

46. Buelens K, Hillmayer K, Compernolle G, Declerck PJ, Gils A: Biochemical Importance of Glycosylation in Thrombin Activatable Fibrinolysis Inhibitor. Circ Res 2007, 102(3):295–301.

47. Zhao L, Morser J, Bajzar L, Nesheim M, Nagashima M: Identification and characterization of two thrombin-activatable fibrinolysis inhibitor isoforms. Thromb Haemost 1998, 80(6):949–955.

48. Knecht W, Willemse J, Stenhamre H, Andersson M, Berntsson P, Furebring C, Harrysson A, Hager AC, Wissing BM, Hendriks D, et al.: Limited mutagenesis increases the stability of human carboxypeptidase U (TAFIa) and demonstrates the importance of CPU stability over proCPU concentration in down-regulating fibrinolysis. Febs J 2006, 273(4):778–792.

49. Ceresa E, De Maeyer M, Jonckheer A, Peeters M, Engelborghs Y, Declerck PJ, Gils A: Comparative evaluation of stable TAFIa variants importance of alpha-helix 9 and beta-sheet 11 for TAFIa (in)stability. J Thromb Haemost 2007, 5(10):2105–12.

50. Brouwers GJ, Vos HL, Leebeek FW, Bulk S, Schneider M, Boffa M, Koschinsky M, van Tilburg NH, Nesheim ME, Bertina RM, et al.: A novel, possibly functional, single nucleotide polymorphism in the coding region of the thrombin-activatable fibrinolysis inhibitor (TAFI) gene is also associated with TAFI levels. Blood 2001, 98(6):1992–1993.

51. Schneider M, Boffa M, Stewart R, Rahman M, Koschinsky M, Nesheim M: Two naturally occurring variants of TAFI (Thr-325 and Ile-325) differ substantially with respect to thermal stability and antifibrinolytic activity of the enzyme. J Biol Chem 2002, 277(2):1021–1030.

52. Marx PF, Havik SR, Bouma BN, Meijers JC: Role of isoleucine residues 182 and 183 in thrombin-activatable fibrinolysis inhibitor. J Thromb Haemost 2005, 3(6):1293–1300.

53. Deutsch DG, Mertz ET: Plasminogen: purification from human plasma by affinity chromatography. Science 1970, 170(962):1095–1096.

54. Bury AF: Evaluation of three sodium dodecyl sulphate-polyacrylamide gel electrophoresis buffer systems. J Chromatogr 1981, 213:491–450.

55. Matsudaira P: Sequence from picomole quantities of proteins electroblotted onto polyvinylidene difluoride membranes. J Biol Chem 1987, 262(21):10035–10038.

56. Hendriks D, Scharpe S, van Sande M: Assay of carboxypeptidase N activity in serum by liquid-chromatographic determination of hippuric acid. Clin Chem 1985, 31(12):1936–1939.

57. Falls LA, Farrell DH: Resistance of gammaA/gamma' fibrin clots to fibrinolysis. J Biol Chem 1997, 272(22):14251–14256.

58. Gobom J, Nordhoff E, Mirgorodskaya E, Ekman R, Roepstorff P: Sample purification and preparation technique based on nano-scale reversed-phase columns for the sensitive analysis of complex peptide mixtures by matrix-assisted laser desorption/ionization mass spectrometry. J Mass Spectrom 1999, 34(2):105–116.

A Novel Method for Screening the Glutathione Transferase Inhibitors

Zhijun Wang, Li Jin, Grzegorz Węgrzyn and Alicja Węgrzyn

ABSTRACT

Background

Glutathione transferases (GSTs) belong to the family of Phase II detoxifica-tion enzymes. GSTs catalyze the conjugation of glutathione to different en-dogenous and exogenous electrophilic compounds. Over-expression of GSTs was demonstrated in a number of different human cancer cells. It has been found that the resistance to many anticancer chemotherapeutics is directly correlated with the over-expression of GSTs. Therefore, it appears to be im-portant to find new GST inhibitors to prevent the resistance of cells to anti-cancer drugs. In order to search for glutathione transferase (GST) inhibitors, a novel method was designed.

Results

Our results showed that two fragments of GST, named F1 peptide (GY-WKIKGLV) and F2 peptide (KWRNKKFELGLEFPNL), can significantly inhibit the GST activity. When these two fragments were compared with several known potent GST inhibitors, the order of inhibition efficiency (measured in reactions with 2,4-dinitrochlorobenzene (CDNB) and glutathione as substrates) was determined as follows: tannic acid > cibacron blue > F2 peptide > hematin > F1 peptide > ethacrynic acid. Moreover, the F1 peptide appeared to be a noncompetitive inhibitor of the GST-catalyzed reaction, while the F2 peptide was determined as a competitive inhibitor of this reaction.

Conclusion

It appears that the F2 peptide can be used as a new potent specific GST inhibitor. It is proposed that the novel method, described in this report, might be useful for screening the inhibitors of not only GST but also other enzymes.

Background

Glutathione transferase (GST) (EC 2.5.1.18) is a multifunctional enzyme, which protects cells against cytotoxic and genotoxic stresses. GST catalyzes the conjugation of cytotoxic agents to glutathione (γ-glutamyl-cysteinyl-glycine), producing less reactive chemical species. Changes in GST levels have been found to correlate with resistance to anticancer drugs through accelerated detoxification of these drugs' substrates [1-4].

Members of the GST family are present at relatively high concentrations in the cytosol of various mammalian tissues. Over-expression of GST isozymes has been reported in a number of different human cancers, when compared to the corresponding normal tissues [5,6]. A 2-fold increase in GST activity was found in lymphocytes from chronic lymphocytic leukemia (CLL) patients, who were resistant to chlorambucil, relative to lymphocytes from untreated CLL patients [7]. As GST isozymes are frequently up-regulated in many solid tumors and lymphomas, inhibition GST activity has become a new drug design concept [8-13]. These facts led to the search for and design of GST inhibitors, including their synthetic analogues and glutathione conjugates, however, most of the existing inhibitors are either too toxic to be used in vivo or are effective only in vitro [14,15].

Although several different GST inhibitors have been reported, to our knowledge, there are no reports on design of the GST inhibitors according to GST sequence. In this report, a novel, covering all gene fragments (CAGF), cloning method was used to screen the GST fragments which can bind to glutathione and form the inhibitory complexes. These inhibitory complexes act as modified substrate inhibitors or substrate homologues to inhibit the GST activity. The method described in this report should be suitable not only for development of novel drugs inhibiting the GST activity, but also for finding effective inhibitors in other enzyme-catalyzed reaction systems.

Results

Screening the GST Inhibitors using the Fragments of GST

The scheme of the 'covering all gene fragments' (CAGF) cloning method is shown in Fig. 1, and the whole screening procedure is shown in Fig. 2. Following five-time panning procedure, as described in the Methods section, 150 positive clones, which can tightly bind to the glutathione Sepharose 4B beads, were picked up from the plates. The typical panning efficiency during each round is shown in Table 1. After five-time panning procedure, the fraction of unbound E. coli cells was significantly decreased, from about 11% to 3.9×10^{-5}%.

Table 1. The binding efficiency of E. coli cells after each round of panning procedure on glutathione Sepharose 4B beads.

E. coli cells	Panning round	Input E. coli cells	Unbound E. coli cells	Elution efficiency (%)
E. coli cell expressing GST fragments	1	3.8×10^{10}	4.2×10^9	11.05
	2	4.2×10^{10}	5.8×10^7	0.14
	3	4.8×10^{10}	5.1×10^6	1.1×10^{-2}
	4	5.6×10^{10}	6.5×10^5	1.2×10^{-3}
	5	6.9×10^{10}	2.7×10^4	3.9×10^{-5}

The 150 positive clones were picked up from the plates and used for screening the GST inhibitors. Following five consecutive screening procedures (consisting of screening the binding of peptides to glutathione Sepharose 4B beads, and screening the positive clones as GST inhibitors), the inhibitor efficiencies of all positive clones were compared. We found that positive clones expressing GYWKIKGLV (F1 peptide) and KWRNKKFELGLEFPNL (F2 peptide) can significantly inhibit GST activity. The binding efficiency of E. coli cells expressing F1 or F2 peptides on the glutathione Sepharose 4B beads was confirmed by an independent experiment. The fraction of E. coli cells expressing F1 or F2 peptides unbound to the glutathione Sepharose 4B beads was 2.3×10^{-5}% or 1.1×10^{-5}%, respectively, while 26.4% control E. coli cells remained unbound to such beads (Table 2).

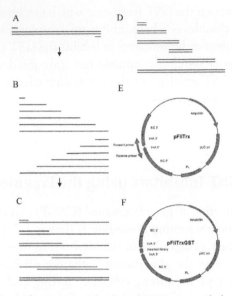

Figure 1. Cloning all GST gene fragments into the plasmid DNA vector with the covering all gene fragments (CAGF) cloning method. A): The gene fragments of GST, B): The amplification of GST fragments using the system containing ddNTP, which can terminate the amplification reaction, and produce the DNA sequences with the single base differences, thus, the reaction system can produce a large library of fragments with single base differences. C): The binding of amplified products, D): Digestion of the CAGF cloning products with Exonuclease VII to form the blunt-ended DNA fragments. E): Amplification of the whole pFliTrx plasmid with the primers FP2 and RP2, F): The linearized pFilTrx plasmid was linked with the DNA library of the gene encoding GST.

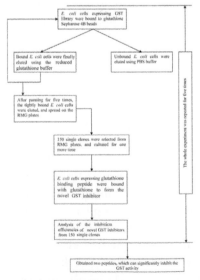

Figure 2. The experimental procedure for screening the fragments of GST which can significantly inhibit GST activity.

Table 2. The binding efficiency of E. coli cells expressing F1 and F2 peptides on glutathione Sepharose 4B beads.

E. coli cells	Input E. coli cells	Unbound E. coli cells	Elution efficiency (%)
E. coli cells expressing F1 peptide	3.0×10^{10}	6.8×10^{4}	2.3×10^{-5}
E. coli cells expressing F2 peptide	2.8×10^{10}	3.2×10^{4}	1.1×10^{-5}
E. coli cells (control)	3.6×10^{10}	9.5×10^{9}	26.4

It has been concluded from the crystallographic study that Arg41, Lys44 and Asn53 of GST can interact with glutathione [16]. Here, our results show that the identified F2 peptide KWRNKKFELGLEFPNL contains Arg41, Lys44 and Asn53 (indicated in bold letters). The control peptide F4, lacking Arg41, Lys44 and Asn53 residues, could not bind efficiently to the glutathione Sepharose 4B beads (Table 3). Moreover, Tyr6, Trp7 and Leu12 of GST were shown to interact with glutathione [16], and our results show that F1 peptide GYWKIKGLV contains Tyr6, Trp7 and Leu12 (indicated in bold letters). Moreover, the control peptide F3, lacking Tyr6, Trp7 and Leu12, could not bind efficiently to glutathione Sepharose 4B beads (Table 3).

Table 3. The binding of synthesized peptides F1, F2, F3 and F4 to glutathione Sepharose 4B beads.

Peptide	Binding efficiency	
	Before elution	After elution
F1	99.6%	2.1%
F2	99.3%	1.8%
F3	1.2%	1.2%
F4	2.6%	2.5%

The binding efficiencies of different peptides were determined according to the ratio of peptide concentrations after the binding and before the binding experiments. Moreover, the peptide binding characteristics were determined again, after the peptides were eluted with the elution buffer.

Results of the screening with the use of the CAGF cloning method are consistent with the crystallographic data. The structure-function analysis has shown that GST contains one important binding site (G-site) for glutathione [1]. Experiments based on kinetic and chemical modification techniques indicated that the active site might contain either His, Cys, Trp, Arg, or Asp [17-21]. The crystal structure indicates that GST binds two molecules of glutathione sulfonate at the G-site. Several groups have investigated changes in amino acids involved in the formation of the G site of GST. The Tyr6, as one of the important components of the G site, is conserved in many mammalian GSTs. Tyr6 plays an essential role in stabilizing the thiolate anion of glutathione through hydrogen bonding. This residue was studied using site-directed mutagenesis, and when Tyr was replaced by different amino acids, GST has lost at least 90% its specific activity [22-24]. Our results with the CAGF cloning method also suggest an important function of Tyr6 in glutathione binding, therefore, the screening results are consistent with the crystallographic data.

The Binding Characteristics of F1 and F2 Peptides

To determine the binding characteristics of selected peptides (F1 and F2), an analysis of the interaction of synthesized peptides with glutathione Sepharose 4B beads was performed using the Scatchard method. The Scatchard analysis is a method of linearizing data from the binding experiment in order to determine binding capacity. The ratio of specific binding and free concentrations was plotted against specific binding concentration. The maximum binding capacity B_{max} and dissociation constant K_d of F1 and F2 peptides were determined. Our results show that there are about 1.1 F1 peptide and 1.2 F2 peptide binding sites on glutathione Sepharose 4B beads, and the disassociation constant of the F2 peptide is lover than that of the F1 peptide (Table 4).

Table 4. The binding of F1 and F2 peptides to glutathione.

Peptide	Binding characteristics of peptides on glutathione Sepharose 4B beads	
	B_{max} (site)	K_d (pM)
F1 peptide	1.1	45.6
F2 peptide	1.2	18.3

The experiments were performed according to the binding characteristics of F1 and F2 peptides to the glutathione Sepharose 4B beads. The concentrations of bound and free peptides were determined following separation of the binding complexes by centrifugation, and the free peptide concentration in the supernatant was detected by the Lowry method.

The binding efficiencies of F1-glutathione and F2-glutathione to GST were further confirmed by analysis of the binding of GST to the F1 peptide-glutathione Sepharose 4B complex and the F2 peptide-glutathione Sepharose 4B complex. The appropriate binding capacity Bmax and dissociation constant Kd values were determined. The results show that there are about 1.2 GST binding sites for F1 peptide-glutathione Sepharose 4B complex, and 1.5 GST binding sites for F2 peptide-glutathione Sepharose 4B complex. The disassociation constant of GST on F2 peptide-glutathione Sepharose 4B complex was lower than that of GST on F1 peptide-glutathione Sepharose 4B complex (Table 5).

Table 5. The binding of F1 and F2 peptides to GST.

Peptide-glutathione Sepharose 4B beads	Binding characteristics of GST on peptide-glutathione Sepharose 4B beads	
	B_{max} (site)	K_d (pM)
F1 peptide-glutathione Sepharose 4B beads	1.2	156.3
F2 peptide-glutathione Sepharose 4B beads	1.5	114.2

The experiments were performed according to the binding characteristics of GST with F1 peptide-glutathione Sepharose 4B bead complex and F2 peptide-glutathione Sepharose 4B bead complex. The concentrations of bound and free GST were determined after separation of complexes by centrifugation, and the free GST concentration in the supernatant was determined by the Lowry method.

The Inhibitory Effects of Selected Peptides

The synthesized peptides F1 and F2 were used in the analysis of the efficiency of inhibition of GST activity (Table 6). The inhibition efficiencies were as follows: tannic acid > cibacron blue > F2 peptide > hematin > F1 peptide > ethacrynic acid, with the use of CDNB or DCNB as the substrates. Moreover, we could not find any significant inhibition of the GST activity using control peptides F3 or F4.

Table 6. Effects of different inhibitors (two selected peptides F1 and F2, tannic acid, cibacron blue, hematin, and ethacrynic) on the GST activity.

Inhibitor (1 µM)	Specific GST activity (units/mg)	
	CDNB (1 mM)	DCNB (1 mM)
Control	2.80 ± 0.01 (100%)	4.51 ± 0.02 (100%)
F1	1.20 ± 0.01(43.1%)	2.65 ± 0.02 (58.8)
F2	0.73 ± 0.01 (26.1%)	1.49 ± 0.02 (33.2%)
tannic acid	0.16 ± 0.01 (5.6%)	0.49 ± 0.01 (10.9%)
cibacron blue	0.51 ± 0.01 (18.3%)	0.96 ± 0.01 (21.3%)
hematin	0.99 ± 0.01 (35.2%)	1.90 ± 0.02 (42.1%)
ethacrynic acid	1.54 ± 0.01 (55.1%)	3.66 ± 0.02 (81.2%)

Experiments were performed using glutathione and CDNB or DCNB as substrates. The activity was determined as described in the Methods section. Mean values ± SD for triplicate determinations are shown. A unit of GST activity was defined as the formation of 1 µmol of the product per min under the conditions of the specific assay. Specific activity is defined as the units of enzyme activity per mg of GST in the reaction system.

These results indicate that we have found an efficient GST inhibitor, the F2 peptide, which is more efficient than hematin (35.2% activity with CDNB as a substrate, 42.1% activity with DCNB as a substrate). We also found another inhibitor of this reaction, the F1 peptide, which is more efficient than ethacrynic acid (55.1% activity with CDNB as a substrate, 81.2% activity with DCNB as a substrate).

The Inhibition Characteristics of Selected Peptides

The inhibitory effects of selected F1 and F2 peptides on the GST-catalyzed reaction, using CDNB and glutathione as substrates, are shown in Fig. 3. Peptides F1 and F2 inhibited the reaction in a dose-dependent manner, with 50% inhibitory concentrations of 0.8 µM and 0.6 µM, respectively.

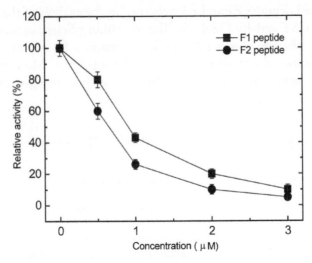

Figure 3. Effects of F1 and F2 peptides on the activity of GST (µmol/mg/min). The activity was measured by monitoring the GST activity with different peptide concentrations, 1 mM glutathione (GSH) and 1 mM CDNB in 100 mM potassium phosphate buffer (pH6.5) at 25°C. Each point shows the mean ± SD of triplicate determinations. Relative GST activity was obtained from the ratio of GST activity in the presence of inhibitor and without inhibitor.

To obtain information on the nature of the inhibition by F1 and F2 peptides, GST activity was measured with variable concentrations of glutathione. Here, we applied a new model in describing enzyme inhibitor (Fig. 4). It is possible that when the F1 or F2 peptide bound to the substrate glutathione, a peptide-glutathione complex was formed. Because the concentration of the peptide inhibitor was significantly lower than the substrate (glutathione) concentration, we assumed that the binding of the peptide inhibitor with glutathione will not significantly affect the substrate concentration. Hence, the newly formed peptide-glutathione complex was considered as a new inhibitor. The concentration of the peptide-glutathione inhibitor is almost equal to the peptide concentration, thus, the Michaelis-Menten equation was used in the analysis.

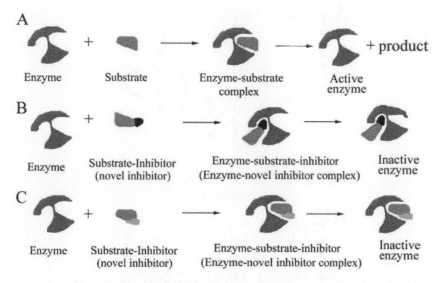

Figure 4. The putative mechanisms of F1 peptide- and F2 peptide-mediated inhibition of the GST activity. A): Binding of the enzyme (blue) and substrate (red) results in formation of the enzyme-substrate complex, and then in liberation of the final product. B): Binding of the F1 peptide (black) and substrate (red) leads to formation of a new peptide-substrate complex inhibitor, which may non-competitively inhibit the enzyme activity. C): Binding of the F2 peptide (green) and substrate (red) leads to formation of a new peptide-substrate complex inhibitor, which may competitively inhibit the enzyme activity.

The Lineweaver-Burk plot for glutathione as the variable peptide concentration was applied to determine the types of inhibition caused by F1 and F2 peptides. The plot provides a useful graphical method for analysis of the Michaelis-Menten equation. The effects of the peptide on GST-catalyzed reaction kinetics were determined by analysis of apparent Vmax, inhibitor constant Ki and [I]/Ki values. Our results indicated that the F1 peptide exerted a noncompetitive inhibition in the GST-catalyzed reaction with the changing glutathione and F1 peptide concentrations (Vmax decreased while Km remained unchanged) (Fig. 5A). However, the F2 peptide exerted a competitive inhibition in this reaction, with the changing glutathione and F2 peptide concentrations (Km decreased while Vmax remained unchanged) (Fig. 5B).

The Vmax, Ki and [I]/Ki values for the F1 peptide were determined (Table 7). The Vmax value of the GST-catalyzed reaction was determined as 1 μmol/mg/min. However, in the presence of 0.8 μM F1 peptide inhibitor, the Vmax, Ki and [I]/Ki values were determined as 0.833 μmol/mg/min, 4.0 μM and 0.2, respectively. With the increasing F1 peptide concentrations, from 0.8 to 3.2 μM, the Vmax value decreased from 0.833 μmol/mg/min to 0.645 μmol/mg/min, Ki value increased from 4.0 μM to 5.82 μM, and [I]/Ki value increased from 0.2 to 0.55.

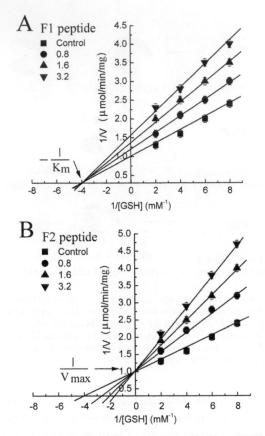

Figure 5. Lineweaver-Burk plot of the GST activity with varying glutathione (GSH) concentrations. A): The activity was measured with 1 mM CDNB, and different concentrations of GSH, and the F1 peptide. B): The activity was measured with 1 mM CDNB, and different concentrations of GSH, and the F2 peptide.

Table 7. The characterization of the F1 peptide as an inhibitor of the GST-catalyzed reaction (The V_{max}, K_i and $[I]/K_i$ values of the GST-catalyzed reaction in the presence of the F1 inhibit peptide were determined).

F1 peptide concentration [I] (µM)	V_{max} value (µmol/mg/min)	K_i value (µM)	$[I]/K_i$
0	1.0	-	-
0.8	0.833	4.0	0.2
1.6	0.741	4.57	0.35
3.2	0.645	5.82	0.55

The Km, Ki and [I]/Ki values for the F2 peptide were also determined (Table 8). The Km value of the GST-catalyzed reaction was determined as 0.25 mM. However, in the presence of 0.8 µM F2 peptide inhibitor, the Km, Ki and [I]/Ki values were determined as 0.38 mM, 1.53 µM and 0.52, respectively. With the increasing F2 peptide concentrations from 0.8 to 3.2 µM, the Km value increased from 0.38 mM to 0.58 mM, Ki value increased from 1.53 μM to 2.42 μM, and [I]/Ki value increased from 0.52 to 1.32.

Table 8. The characterization of the F2 peptide as an inhibitor of the GST-catalyzed reaction (The V_{max}, K_i and [I]/K_i values of the GST-catalyzed reaction in the presence of the F2 inhibit peptide were determined).

F2 peptide concentration [I] (μM)	K_m value (mM)	K_i value (μM)	[I]/K_i
0	0.25	-	-
0.8	0.38	1.53	0.52
1.6	0.50	1.60	1.0
3.2	0.58	2.42	1.32

Moreover, with the changing concentrations of CDNB or DCNB, from 0.5 mM to 2 mM, the kinetics of the GST-catalyzed reaction remained similar in the reaction system containing GST, glutathione and the inhibitor (F1 peptide or F2 peptide). Therefore, we conclude that CDNB and DCNB cannot significantly affect the inhibition efficiency of F1 peptide or F2 peptide.

All these results show that effective non-competitive inhibitor F1 peptide and competitive inhibitor F2 peptide were found by using the CAGF cloning method. Although F1 and F2 peptides comprise only a small part of GST, they show significant inhibition efficiencies in the GST-catalyzed reaction.

Discussion

The development of resistance to anticancer agents is a primary concern in cancer chemotherapy. In this light, it is obvious that the emergence of drugs, such as the GST inhibitors, able to overcome this resistance is a advancement [10,11]. Therefore, it is of special interest to develop GST inhibitors able to enhance the therapeutic index of anticancer drugs. Ethacrynic acid and quinine, which are both GST inhibitors, have been reported to reverse the resistance to melphalan and doxorubicin of cancer cell lines with increased GST expression [25]. In fact, ethacrynic acid has been used as an inhibitor of GST in vivo. However, first-generation GST inhibitors (e.g. ethacrynic acid) were unsuccessful in clinical trials. This might be due to its lack of specific function for GST isozyme, and propensity to react with other chemicals. In addition, there caused a number of unwanted clinical side effects. Therefore, more specific GST inhibitors may eliminate some of these undesirable features.

Here, we used the CAGF cloning method to find the GST fragments interacting with glutathione, which might be useful for the finding of GST inhibitors. We found two inhibitory peptide fragments, F1 peptide and F2 peptide. Our results revealed that F2 peptide is a potent inhibitor of the reaction with IC50 of 0.6 µM (Fig. 3).

The putative inhibition mechanisms of actions of F1 and F2 peptide inhibitors are shown in Fig. 4. The F1-glutathione complex was found to be a non-competitive inhibitor, suggesting that the inhibitory binding site of F1-glutathione is different from the catalytic site of GST. An analysis of the crystal structure of GST and glutathione shows that the F1 peptide is located in the interior position of the G site (Fig. 6A). It may be difficult for F1-glutathione complex to dock into the catalytic site of GST. We assume that F1-glutathione complex may be docked into non-catalytic site of GST, affect the structure of catalytic site of GST, and cause the non-competitive inhibition.

On the other hand, the F2 peptide is located in the marginal position of the G site of GST (Fig. 6B). It may be easy for F2-glutathione complex to duck into the catalytic site of GST. We assume that the F2-glutathione complex may be docked into the G site of GST. Thus, F2-glutathione may directly interfere with the catalytic site of GST and glutathione. Since the F2-glutathione causes a competitive inhibition, the F2 peptide may be a good candidate for further studies on cancer chemomodulation.

A

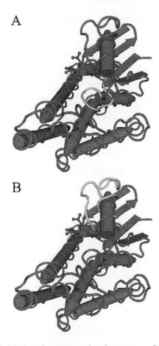

B

Figure 6. The crystal structure of GST with putative binding sites of F1 peptide (A) and F2 peptide (B) inhibitors. Symbols: GST (purple), glutathione (brown), selected peptide (yellow). The crystal structure of GST and glutathione complex was viewed using the protein structure data (PDB: 1m99).

The following mechanism was used to explain the inhibitory activity of GST fragment-substrate complexes on GST-catalyzed reaction. When the F1 or F2 peptide bound to the substrate glutathione, a peptide-glutathione complex was formed. Although GST can convert glutathione into the reaction products, this enzyme cannot convert the peptide-glutathione inhibitor into the product. Thus, the binding of the peptide-glutathione to GST can inactive the enzyme activity. We speculate that peptide-glutathione occupied the functional domain or affected the functional domain of GST. Thus, GST-peptide-glutathione or GST-glutathione-peptide complex cannot catalyze the conversion of glutathione (Fig 4B and 4C). Here, the function of peptide-glutathione inhibitor is just like the substrate homologue or substrate-modifying inhibitor [26].

In summary, we have determined two glutathione-binding fragments of the GST sequence, and found that the F2 peptide, selected by the CAGF cloning method, can be considered as the inhibitor of GST. The F2 peptide is a potent inhibitor, stronger than hematin and ethacrynic acid, but weaker than tannic acid and cibacron blue. We suggest that the F2 peptide can be considered in applications against GST-induced multidrug resistance.

Moreover, Fig. 7 shows the scheme of the novel method in finding enzyme inhibitors. In the first step, the enzyme fragments which can bind with the substrate were screened to find the binding peptides. In the second step, the complexes of enzyme fragments and substrate were screened to find the enzyme inhibitor. We believe that this method can be used as a common tool for finding enzyme fragments that interact with a substrate, and subsequently for finding enzyme inhibitors.

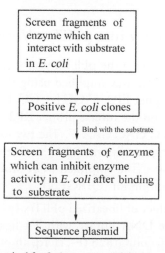

Figure 7. The scheme of the novel method for finding enzyme inhibitors.

Conclusion

In conclusion, we have successfully found a F2 peptide as GST inhibitor with the novel screening method from GST sequence. Our screening method should be useful for screening many different enzyme inhibitors.

Methods

Generation of the GST Library

The forward primer FP1: 5' ATG TCC CCT ATA CTA GGT 3' and reverse primer RP1: 5' TCA CGA TGC GGC CGC TCG 3' were used to amplify the Schistosoma japonicum full-length GST gene [27] from the pGEX4T-2 plasmid DNA vector (Amersham). The amplified DNA fragments were purified with the QIAquick PCR purification kit (QIAGEN). The following reaction system was used: 1 ng pGEX4T-2 plasmid, 50 mM Tris-HCl (pH 7.8), 5 mM MgCl2, 10 mM 2-mercaptoethanol, 10 µg/ml BSA, 1 ng forward primer FP1 and 1 ng reverse primer RP1, 20 µM dNTP, 1 µM dideoxynucleotides (ddNTP), 2 units DNA Polymerase I (Invitrogen), and ddH2O to the reaction volume of 100 µl; incubation at 15°C for 30 to 60 min.

The amplified DNA library was purified by using the phenol-chloroform method, and dissolved in water. The DNA library was digested with the Exonuclease VII (Epicentre), which has a high enzymatic specificity for single-stranded DNA and exhibits both 5' → 3' and 3' → 5' exonuclease activities. This enzyme is especially useful for rapid removal of single-stranded oligonucleotide primers.

Cloning the GST Library into pFliTrx Vector

The cloning of DNA library into the pFliTrx vector (Invitrogen) was performed as shown in Fig. 1. The pFliTrx was amplified using Pfx DNA polymerase (Invitrogen) with the forward primer FP2: 5' GGT CCG TCG AAA ATG ATC GCC CCG ATT CTG GAT 3' and the reverse primer RP2: 5' CGG ACC GCA CCA CTC TGC CCA GAA ATC GAC GAA 3'. The two-step PCR reaction was performed under the following conditions: 92°C for 2 min; then 35 cycles at 68°C for 5 min, and 92°C for 30 s. The amplified PCR product was purified by using the QIAquick PCR purification kit (QIAGEN).

The purified PCR product of linearized pFliTrx (without the fusion junction) was used to link it to the DNA library with T4 ligase. The ligation products were introduced into the E. coli GI826 (F-, lacIq, ampC::Ptrp::cI, ΔfliC, ΔmotB, eda::Tn10) (Invitrogen).

Screening the GST Fragments which can Bind to Glutathione

The GST library was introduced to E. coli GI826 competent cells, which were then cultured in 50 ml of IMC medium (1 × M9 salts, 0.2% casamino acid, 0.5% glucose, 1 mM $MgCl_2$) containing 100 µg/ml ampicillin with shaking (225–250 rpm) to saturation (OD_{600} = 3) for 15 hours at 25°C. E. coli cells were added to 50 ml IMC medium containing 100 µg/ml ampicillin and 100 µg/ml tryptophan for induction. The cells were grown at 25°C with shaking for 6 hours. Then, 1 ml of glutathione Sepharose 4B (Amersham) slurry and 1 ml of tryptophan-induced culture broth were added to 40 ml of the PBS buffer in a 50 ml tube, and kept at the room temperature for 30 min, centrifuged at 1,000 × g for 10 min at the room temperature, then resuspended in the PBS buffer, and centrifuged at 1,000 × g for three more times. Finally, the pellet was resuspended in 2 ml of PBS, and 500 µl of elution buffer (50 mM Tris-HCl, 10 mM reduced glutathione, pH 8.0) were added to elute the bound E. coli cells. The eluted E. coli cells were used for the next panning procedure. Following the panning procedure, 100 µl of the eluted solution was added on the RMG plates (1 × M9 salt, 0.2% casamino acid, 0.5% glucose, 1 mM $MgCl_2$, 1.5% agar). The plates contained 100 µg/ml ampicillin for selection of the positive clones. Then the single positive clones from the RMG plates were picked up, and 150 single clones were used for screening the GST inhibitors.

Screening of GST Fragments which can Inhibit the GST Activity

150 single positive clones (grown on the RGB plates), that could tightly bind to the glutathione Sepharose 4B, were picked up from the plates, and cultured separately in 50 ml of IMC medium containing 100 µg/ml ampicillin for 15 hours at 25°C, then induced with 100 µg/ml tryptophan for 6 hours. E. coli cells were washed with the PBS buffer for three times at 4°C, and suspended in the 5 ml PBS solution, respectively.

Recombinant S. japonicum GST, glutathione, tannic acid, cibacron blue, hematin, ethacrynic acid, 1,2-dichloro-4-nitrobenzene (DCNB) and 2,4-dinitrochlorobenzene (CDNB) were purchased from Sigma-Aldrich, and used to measure the GST activity.

To measure the GST activity, glutathione and CDNB solutions were added (to final concentration of 1 mM) to 100 µl of E. coli cell suspension (108 cells). Then, GST solution was added (the cell suspension without glutathione was used as the control). The GST activity was measured by using the spectrophotometric assay [28].

The single clones, which can produce the GST inhibitory peptide, were selected again, and the plasmid DNA was extracted for determination of inserted sequences. The whole screening procedures were performed five times. Finally, plasmid DNAs were extracted from E. coli cells expressing inhibitory peptides, and sequenced.

Analysis of the Binding of Peptides to Glutathione

The binding characteristics of selected peptides to glutathione were determined according to the analysis of binding of synthesized peptides to the glutathione Sepharose 4B beads. The amount of glutathione in the glutathione Sepharose 4B beads was estimated according to the assumption (according to the manufacturere's information) that there are about 200–400 μmol glutathione/g dried beads. An average value of 300 μmol glutathione/g dried beads was used to calculate the amount of glutathione in the glutathione Sepharose 4B beads. In our experiments, appropriate amount of wet glutathione Sepharose 4B beads (equal to 1 mg dry beads) was added to a 1.5 ml Eppendorf tube, and different amounts of synthesized peptides were added into the tube. After binding for 10 min at 37°C, the binding complexes were separated by centrifugation (12,000 × g for 10 min) and concentrations of bound and free peptides were determined by using the Lowry method [29]. Scatchard analysis was used to determine the K_d and B_{max} values. B_{max} means the maximum binding sites of synthesized peptide on glutathione Speharose 4B beads (μmol peptide/μmol glutathione). K_d is a dissociation constant (pM). Thus, a low K_d value indicates a high affinity.

Analysis of the Binding of Peptide-Glutathione Complex to GST

The binding of selected peptide-glutathione complexes with GST were determined on the basis of analysis of binding of GST to the peptide-glutathione Sepharose 4B bead complex. Wet glutathione Sepharose 4B beads (equal to 1 mg dry wet) was added into a 1.5 ml Eppendorf tube for binding to peptides. After the binding of peptides to glutathione Sepharose 4B beads at 37°C for 10 min, the unbound peptides were washed out. Then, different amounts of GST were added into the tube. After binding for 10 min at 37°C and separation of the bound complexes by centrifugation, the amounts of bound and free GST were determined by the Lowry method [29]. Scatchard analysis was used to determine the K_d and B_{max} values. B_{max} means the maximum binding site of GST with the peptide on peptide-glutathione Speharose 4B beads (μmol GST/μmol peptide-glutathione). K_d is a disassociation constant (pM).

Enzyme Inhibition Assay

When the screening experiments were performed, four peptides were synthesized to analyze their inhibition efficiencies. F1 peptide (GYWKIKGLV, yield: 25.3 mg), F2 peptide (KWRNKKFELGLEFPNL, yield: 28.1 mg), F3 peptide (GKIKGV, yield: 2.2 mg) and F4 peptide (KWNKFELGLEFPL, yield: 1.9 mg) were obtained from the Invitrogen (Custom Peptide Synthesis, with the purity > 95%). Recombinant S. japonicum GST (~40 units/mg), glutathione, tannic acid, cibacron blue, hematin, ethacrynic acid, 1,2-dichloro-4-nitrobenzene (DCNB) and 2,4-dinitrochlorobenzene (CDNB) were from Sigma-Aldrich Co. The inhibition studies were carried out according to the previously described method [28] at 25°C using glutathione (1 mM) and CDNB (1 mM) or DCNB (1 mM) as substrates. The inhibitors (tannic acid, cibacron blue, hematin, ethacrynic acid or the synthesized peptides) were added to the reaction mixture and GST activity was determined.

The peptide concentration resulting in 50% inhibition (IC50) was determined from a plot of remaining activity versus peptide concentration. Protein concentration was measured according to the Lowry method [29]. Enzyme inhibitory kinetic studies were carried out using various concentrations of glutathione and CDNB and different concentrations of synthetic peptides (0.8, 1.6, 3.2 µM).

Abbreviations

CAGF: covering all gene fragments; GST: glutathione transferase; CDNB: 2,4-dinitrochlorobenzene; DCNB: 1,2-dichloro-4-nitrobenzene; CLL: chronic lymphocytic leukemia.

Competing Interests

The authors declare that they have no competing interests.

Authors' Contributions

All authors have contributed to the content of the article, and all authors approved the final version.

Acknowledgements

This work was supported by National Natural Science Foundation of China (Grant number: 30400077), University of Gdańsk (task grant no. DS/1480-4-114-09) and Institute of Biochemistry and Biophysics of the Polish Academy of Sciences (task grant 32.1).

References

1. Mannervik B, Danielson UH: Glutathione transferases-structure and catalytic activity. CRC Crit Rev Biochem 1988, 23:283–337.

2. Pickett CB, Lu AY: Glutathione S-transferases: gene structure, regulation, and biological function. Annu Rev Biochem 1989, 58:743–764.

3. Coles B, Ketterer B: The role of glutathione and glutathione transferases in chemical carcinogenesis. Crit Rev Biochem Mol Biol 1990, 25:47–70.

4. Douglas KT: Mechanism of action of glutathione-dependent enzymes. Adv Enzymol Relat Areas Mol Biol 1987, 59:103–167.

5. Adang AE, Brussee J, Gen A, Mulder GJ: The glutathione-binding site in glutathione S-transferases. Investigation of the cysteinyl, glycyl and gamma-glutamyl domains. Biochem J 1990, 269:47–54.

6. Abramovitz M, Homma H, Ishigaki S, Tansey F, Cammer W, Listowsky I: Characterization and localization of glutathione-S-transferases in rat brain and binding of hormones, neurotransmitters, and drugs. J Neurochem 1988, 50:50–57.

7. Schisselbauer JC, Silber R, Papadopoulos E, Abrams K, LaCreta FP, Tew KD: Characterization of glutathione S-transferase expression in lymphocytes from chronic lymphocytic leukemia patients. Cancer Res 1990, 50:3562–3568.

8. Wilce MC, Parker MW: Structure and function of glutathione S-transferases. Biochim Biophys Acta 1994, 1205:1–18.

9. Wilce MC, Feil SC, Board PG, Parker MW: Crystallization and preliminary X-ray diffraction studies of a glutathione S-transferase from the Australian sheep blowfly, Lucilia cuprina. J Mol Biol 1994, 236:1407–1409.

10. Hayes JD, Pulford DJ: The glutathione S-transferase supergene family: regulation of GST and the contribution of the isozymes to cancer chemoprotection and drug resistance. Crit Rev Biochem Mol Biol 1995, 30:445–600.

11. Tidefelt U, Elmhorn-Rosenborg A, Paul C, Hao XY, Mannervik B, Eriksson LC: Expression of glutathione transferase p as a predictor for treatment

results at differentstages of acute nonlymphoblastic leukemia. Cancer Res 1992, 52:3281–3285.

12. Black SM, Beggs JD, Hayes JD, Bartoszek A, Muramatsu M, Sakai M, Wolf CR: Expression of human glutathione S-transferases in Saccharomyces cerevisiae confers resistance to the anticancer drugs adriamycin andglutachlorambucil. Biochem J 1990, 268:309–315.

13. Dirven HAAM, van Ommen B, van Bladeren PJ: Involvement of human glutathione S-transferase isoenzymes in the conjugation of cyclophosphamide metabolites with glutathione. Cancer Res 1994, 54:6215–6220.

14. Hayeshi R, Chinyanga F, Chengedza S, Mukanganyama S: Inhibition of human glutathione transferases by multidrug resistance chemomodulators in vitro. J Enzyme Inhib Med Chem 2006, 21:581–587.

15. Muleya V, Hayeshi R, Ranson H, Abegaz B, Bezabih MT, Robert M, Ngadjui BT, Ngandeu F, Mukanganyama S: Modulation of Anopheles gambiae Epsilon glutathione transferase activity by plant natural products in vitro. J Enzyme Inhib Med Chem 2008, 23:391–399.

16. Kursula I, Heape AM, Kursula P: Crystal structure of non-fused glutathione S-transferase from Schistosoma japonicum in complex with glutathione. Protein Pept Lett 2005, 12:709–712.

17. Ricci G, Del Boccio G, Pennelli A, Lo Bello M, Petruzzelli R, Caccuri AM, Barra D, Federici G: Redox forms of human placenta glutathione transferase. J Biol Chem 1991, 266:21409–21415.

18. van Ommen B, den Besten C, Rutten AL, Ploemen JH, Vos RM, Muller F, van Bladeren PJ: Active site-directed irreversible inhibition of glutathione S-transferases by the glutathione conjugate of tetrachloro-1,4-benzoquinone. J Biol Chem 1988, 263:12939–12942.

19. Awasthi YC, Bhatnagar A, Singh SV: Evidence for the involvement of histidine at the active site of glutathione S-transferase psi from human liver. Biochem Biophys Res Commun 1987, 143:965–970.

20. Tamai K, Satoh K, Tsuchida S, Hatayama I, Maki T, Sato K: Specific inactivation of glutathione S-transferases in class Pi by SH-modifiers. Biochem Biophys Res Commun 1990, 167:331–338.

21. Chang LH, Wang LY, Tam MF: The single cysteine residue on an alpha family chick liver glutathione S-transferase CL 3-3 is not functionally important. Biochem Biophys Res Commun 1991, 180:323–328.

22. Manoharan TH, Gulick AM, Puchalski RB, Servais AL, Fahl WE: Structural studies on human glutathione S-transferase pi. Substitution mutations to

determine amino acids necessary for binding glutathione. J Biol Chem 1992, 267:18940–18945.

23. Wang RW, Newton DJ, Huskey SE, McKeever BM, Pickett CB, Lu AY: Site-directed mutagenesis of glutathione S-transferase YaYa. Important roles of tyrosine 9 and aspartic acid 101 in catalysis. J Biol Chem 1992, 267:19866–19871.

24. Kolm RH, Sroga GE, Mannervik B: Participation of the phenolic hydroxyl group of Tyr-8 in the catalytic mechanism of human glutathione transferase P1–1. Biochem J 1992, 285:537–540.

25. Hansson J, Berhane K, Castro VM, Jungnelius U, Mannervik B, Ringborg U: Sensitization of human melanoma cells to the cytotoxic effect of melphalan by the glutathione transferase inhibitor ethacrynic acid. Cancer Res 1991, 51: 94–98.

26. Burg D, Mulder GJ: Glutathione conjugates and their synthetic derivatives as inhibitors of glutathione-dependent enzymes involved in cancer and drug resistance. Drug Metab Rev 2002, 34:821–863.

27. Smith DB, Johnson KS: Single-step purification of polypeptides expressed in Escherichia coli as fusions with glutathione S-transferase. Gene 1988, 67:31–40.

28. Habig WH, Pabst MJ, Jakoby WB: Glutathione S-transferases. The first enzymatic step in mercapturic acid formation. J Biol Chem 1974, 249:7130–7139.

29. Lowry OH, Rosebrough NJ, Farr AL, Randall RJ: Protein measurement with the Folin phenol reagent. J Biol Chem 1951, 193:265–275.

Role of the UPS in Liddle Syndrome

Daniela Rotin

ABSTRACT

Hypertension is a serious medical problem affecting a large population worldwide. Liddle syndrome is a hereditary form of early onset hypertension caused by mutations in the epithelial Na⁺ channel (ENaC). The mutated region, called the PY (Pro-Pro-x-Tyr) motif, serves as a binding site for Nedd4-2, an E3 ubiquitin ligase from the HECT family. Nedd4-2 binds the ENaC PY motif via its WW domains, normally leading to ENaC ubiquitylation and endocytosis, reducing the number of active channels at the plasma membrane. In Liddle syndrome, this endocytosis is impaired due to the inability of the mutated PY motif in ENaC to properly bind Nedd4-2. This leads to accumulation of active channels at the cell surface and increased Na⁺ (and fluid) absorption in the distal nephron, resulting in elevated blood volume and blood pressure. Small molecules/compounds that destabilize cell surface ENaC, or enhance Nedd4-2 activity in the kidney, could potentially serve to alleviate hypertension.

Protein Pathway Involvement in Disease

Introduction

Elevated arterial blood pressure, or hypertension, poses a serious public health problem, affecting approximately 25% of the adult population in the industrial world [1], and becoming, along with obesity, a serious health problem in the developing world as well. In recent years, the causes of several genetic disorders leading to hypertension or hypotension have been identified, and deleterious mutations have been mapped to components of the aldosterone pathway, as well as to key ion channels and transporters expressed along the nephron. Prominent examples include Bartter syndrome type I, II or III, Gitelman syndrome, pseudo-hypoaldosteronism I (PHAI) and Liddle syndrome [2]; the latter two are associated with mutations in the epithelial Na+ channel (ENaC) and are discussed in this review, with a particular focus on Liddle syndrome, a hereditary form of hypertension.

The Epithelial Na$^+$ Channel

The amiloride-sensitive ENaC is an ion channel expressed in Na$^+$-transporting epithelia such as those present in the distal nephron, respiratory epithelium, distal colon and taste buds [3]. In the kidney, it is primarily expressed in the distal connecting tubules (CNTs) and cortical collecting tubules (CCTs) of the nephron [4,5], where it provides the rate limiting step for Na$^+$ (and fluid) reabsorption into the blood stream [3,6,7]. This regulation of Na$^+$ and fluid absorption is tightly controlled by the hormones aldosterone (i.e. the renin-angiotensin-aldosterone pathway) and vasopressin (antidiuretic hormone, ADH), which stimulate channel activity [6,8]. The single channel characteristics of ENaC reveal high selectivity for Na$^+$ over K$^+$, low single channel conductance (~5 pS), high sensitivity to amiloride (~100 nM) and slow gating [6]. ENaC activity is primarily regulated by control of its opening (Po) and numbers at the plasma membrane [8].

ENaC is comprised of three subunits, α, β and γ [9], each consisting of two transmembrane domains flanked by a large extracellular loop and two intracellular N- and C-termini, and is preferentially assembled at a stoichiometry of α2βγ [10,11] (although other configurations have been proposed [12,13]). Maximal channel activity is obtained when all three subunits are expressed together, but expression of α alone, or a combination of αβ, or αγ, results in low or moderate channel activity, respectively [9].

Genetic disease-causing mutations in ENaC, as well as mouse models, have shed important light on ENaC function and the pathology of ENaC-related diseases.

For example, loss of function mutations in either α, β, or γ ENaC cause PHAI [14,15], a salt-wasting disease leading to hypotension, which is also mimicked in knockout mouse models lacking β or γ ENaC [16,17], or models expressing reduced levels of α ENaC [18], all of which exhibit reduced levels of channel expression and activity [9,19]. In contrast to the ENaC loss of function mutations causing PHAI, gain of function mutations in this channel cause Liddle syndrome.

Liddle Syndrome

Liddle syndrome (pseudoaldosteronism, OMIM 177200) is an autosomal dominant disease leading to early onset of hypertension. It is associated with hypokalemic alkalosis, reduced plasma rennin activity and low plasma aldosterone levels [20]. Over the past 12 years, work from Lifton's group (Yale University) and others has identified several deletions/mutations that cause Liddle syndrome, all of which map to β or γ ENaC and lead to elevated channel numbers and activity at the plasma membrane, as assessed by heterologously expressing these mutant ENaCs in Xenopus oocytes or cultured mammalian cells [2,21,22]. These genetic defects either delete the C-terminus of β or γ ENaC [23,24], or mutate a proline or a tyrosine within a short sequence, called the PY (Pro-Pro-x-Tyr) motif [25-28]. The PY motif, or extended PY motif (PPxYxxL [21,29]), is highly conserved in the C-termini of all ENaC subunits [26], and serves as a binding site for the Nedd4 family of ubiquitin ligases [30], as assessed by in vitro binding, yeast two-hybrid and co-immunoprecipitation assays, as well as by structural analysis (e.g. [29,30]).

Regulation of ENaC by the Ubiquitin System and its Impairment in Liddle Syndrome

Ubiquitylation, carried out by the sequential activity of E1, E2 and E3 (ubiquitin ligase) enzymes, usually regulates stability of target proteins that are tagged with ubiquitin by the E3 ligases [31]. Most of these proteins are degraded by the proteasome [32]. Recent studies have demonstrated, however, that ubiquitylation of transmembrane proteins can tag them for endocytosis and/or vesicular sorting, often resulting in their degradation in the lysosome [33,34]. This is usually achieved by the presence of ubiquitin binding motifs or domains (e.g. UIM, UBA, CUE, GAT, UEV, VHS) within proteins such as epsin/Eps15, Hrs and GGA, which function to recognize the ubiquitylated transmembrane proteins and facilitate their endocytosis or sorting [35].

Nedd4 family members are E3 ubiquitin ligases that comprise a C2 domain responsible for membrane targeting [36,37], three to four WW domains that bind the PY motifs of ENaC [29,30,38-42], and a ubiquitin ligase HECT (homologous to E6AP carboxyl-terminus) domain [43,44] (Figures 1 and 2). Of the two closely related Nedd4 members, Nedd4-1 and Nedd4-2, the latter binds ENaC more strongly due to the presence of an additional, high affinity WW domain (WW3, out of four WW domains) [41,42]. Accordingly, Nedd4-2 was shown to effectively suppress ENaC activity by enhancing removal of the channel from the plasma membrane [45-47], and ubiquitylation of ENaC was demonstrated to destabilize cell surface ENaC [48] (Figure 2). Indeed, our recent work has demonstrated that Nedd4-2 can ubiquitylate ENaC present at the apical membrane of cultured kidney epithelial cells [49]. The few Nedd4-1 proteins that also contain this high affinity WW3 domain (e.g. human and Drosophila Nedd4-1) are also able to suppress ENaC activity when heterologously expressed in Xenopus oocytes or cultured cells, although in some cases this is prevented in the presence of the C2 domain (for example, in the case of human Nedd4-1 [41,47,50]), possibly (albeit speculatively) due to inhibitory interactions between the C2 and HECT domains.

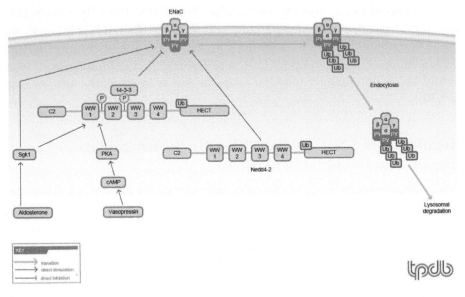

Figure 1. Regulation of ENaC by Nedd4-2 in homeostasis. The ubiquitin ligase Nedd4-2 binds (via its WW domains) to the PY motifs of ENaC, in turn ubiquitylating and targeting ENaC for endocytosis and lysosomal degradation. This process can be inhibited by Sgk1- or Akt-mediated phosphorylation of Nedd4-2, which leads to binding of 14-3-3 proteins to phosphorylated Nedd4-2, thus preventing Nedd4-2 from associating with ENaC and thus increasing ENaC levels at the plasma membrane. Sgk1 can also upregulate ENaC independently of Nedd4-2.

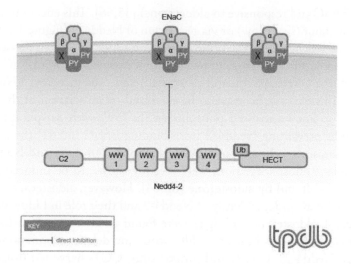

Figure 2. Regulation of ENaC by Nedd4-2 and its impairment in Liddle syndrome. In Liddle syndrome, deletion/mutation of the PY motif in βENaC (or γENaC, not shown) impairs the ability of Nedd4-2 to bind (and thus ubiquitylate) ENaC, leading to accumulation of ENaC channels at the plasma membrane and increased channel activity. (Modified with permission from Staub and Rotin).

Experiments performed in Xenopus oocytes or mammalian cultured cells that ectopically express ENaC reveal that unlike the ability of Nedd4-2 to induce removal of wild-type ENaC from the plasma membrane by ubiquitylation (likely linked to subsequent clathrin-mediated endocytosis [51]), Liddle syndrome mutations in the PY motif of β or γ ENaC severely attenuate this removal, leading to increased retention of mutant ENaC at the cell surface and to elevated channel activity [45,46,49] (Figure 2). Accordingly, conditional knockout mice bearing a deletion of the PY motif in βENaC (a mouse model for Liddle syndrome) develop hypertension that is induced by high salt diet [52]. Moreover, channel feedback inhibition by elevated intracellular Na+ concentrations exhibited by wild-type ENaC (demonstrated ex vivo in the CCTs of rats [53]) is defective in ENaC bearing the Liddle syndrome mutations in the PY motif (as shown by ectopic expression of PY motif-mutated ENaC in Xenopus oocytes [54]), further exacerbating Na+ loading. Together, this results in increased Na+ and fluid reabsorption in the distal nephron, and increased blood volume and blood pressure, which are hallmarks of Liddle syndrome.

As indicated in the section "The epithelial Na+channel," ENaC is tightly (positively) regulated in the kidney by the mineralocorticoid hormone aldosterone. One of the recently discovered aldosterone targets is Sgk1, a Ser/Thr kinase from the Akt (PKB) family that was found to elevate ENaC levels/activity in response to aldosterone in rat or mouse kidney and in A6 cells (a Xenopus cell line endogenously expressing ENaC and responsive to aldosterone) [55,56]. This effect can be mediated either without (see below) or via regulation of Nedd4-2 (Figure 1). Nedd4-2 (but not Nedd4-1) possesses Sgk1 phosphorylation sites and, when phosphorylated by Sgk1 (in cultured cells), is prevented from downregulating ENaC, leading to increased ENaC retention at the cell surface and thus increased ENaC activity [57,58]. This effect is believed to be mediated by association of the adaptor protein 14-3-3, known to bind phosphoSer/Thr [59], with Ser-phosphorylated Nedd4-2, thus preventing Nedd4-2 from binding ENaC [60,61] (by as yet unknown mechanism(s)). In support, the expression of 14-3-3β and Nedd4-2, as well as Nedd4-2 phosphorylation, were recently shown to be induced in CCT cells by dietary salt and by aldosterone [62-64]. However, aldosterone and Sgk1 can stimulate ENaC independently of Nedd4-2 and their role in Liddle syndrome is controversial: aldosterone and Sgk1 were found to increase cell surface abundance of ENaC channels bearing Liddle syndrome deletions/mutations (which cannot bind Nedd4-2) [65-68] and, importantly, CCTs harvested from mutant mice bearing a Liddle syndrome deletion [52] revealed a normal response to aldosterone [69].

In addition to aldosterone, the hormone vasopressin also increases ENaC activity (as well as water absorption) in the distal nephron by binding to V2 receptors and stimulating activation of adenylate cyclase and the release of cAMP [6]. cAMP increases the density of ENaC channels (endogenously or ectopically expressed in epithelial cells) at the plasma membrane [70,71], an effect suggested to be impaired in channels bearing the Liddle syndrome PY motif mutations due to defective trafficking to the cell surface [72], or mobilization from a sub-apical pool [49]. Recent studies also suggested that Nedd4-2 phosphorylation by PKA (which is activated by cAMP) provides inhibitory function much like Sgk1, thus inhibiting the ability of Nedd4-2 to suppress ENaC, leading to increased cell surface abundance of this channel [73].

Among the other factors that regulate ENaC (aside from hormones, ions and Nedd4-2), are proteases such as CAP proteins and TMPRSS3, which activate ENaC by proteolytic cleavage of its ectodomains [3,74-77]. A recent paper suggests that Liddle syndrome mutations increase the number of cleaved (active) ENaCs at the cell surface (thus further increasing Na+ absorption), and that Nedd4-2 and ENaC ubiquitylation regulate the number of cleaved channels at the plasma membrane [78].

Disease Models, Knockouts and Assays

To date, only one mouse model for Liddle syndrome has been generated. These mice, created by the Rossier/Hummler groups (Institut de Pharmacologie et de

Toxicologie, Switzerland) in 1999, bear a deletion of the PY motif in βENaC (a stop codon is inserted at a residue corresponding to residue Arg566 in human βENaC, as found in the original pedigree described by R. Lifton [23]). The mice have no phenotype under a normal salt diet, but develop hypertension when fed a high salt diet [52]. To date, no other relevant knockout mice of Liddle syndrome have been developed.

Disease Targets and Ligands

Liddle syndrome patients are treated with the ENaC antagonist amiloride-tri-amterene and a low salt diet to stabilize their high blood pressure. While Liddle syndrome is a rare disorder, as are several genetic forms of hypertension [2], other forms of hypertension are very common in the population and have no known genetic components. Inhibiting ENaC activity, the rate limiting step in the regulation of Na^+ and fluid reabsorption in the nephron, could provide an attractive target to treat hypertension. With the advent of high throughput technology it is possible to test for inhibition of ENaC activity by screening with small molecules/compound libraries, with the hope of identifying inhibitory compounds that may be superior to amiloride and its analogs. In that regard, we have recently developed a high throughput assay that allows quantification of the amounts of cell surface ENaC (Chen and Rotin, unpublished). Given the key role played by the ubiquitin system/Nedd4-2 in regulating ENaC cell surface stability and ENaC function, identifying compounds that destabilize/decrease ENaC levels at the plasma membrane could have potential therapeutic benefits for the treatment of hypertension. Stimulating Nedd4-2 activity, which leads to ENaC endocytosis/degradation, could also be a possibility. However, since Nedd4-2 likely has other targets in other tissues/cells, this approach needs to be scrutinized to ensure it is targeted specifically to ENaC in the kidney. It is likely that use of putative compounds that aim to enhance ENaC internalization or Nedd4-2 activity would be more effective towards other forms of hypertension and not Liddle syndrome, since the latter carries mutations that already inhibit ENaC internalization and are insensitive to Nedd4-2.

New Frontiers in Drug Discovery

Despite significant recent advances, several key questions remain to be answered regarding the regulation of ENaC by Nedd4-2 and the ubiquitin system. These include:

(i) How is sensing of elevation of intracellular concentrations of Na^+ (that normally shuts down ENaC) related to the Liddle syndrome mutations? If this is regulated via Nedd4-2, how is Nedd4-2 (directly or indirectly) able to sense Na⁺?

(ii) How does phosphorylation of Nedd4-2 by Sgk1 (on sites not within the WW domains), which leads to binding of 14-3-3 to Nedd4-2, inhibit the association of Nedd4-2 with ENaC, which is mediated via the WW domains?

(iii) What is the exact stoichiometry of Nedd4-2-ENaC interactions, and how is it that loss of only one PY motif is sufficient to cause Liddle syndrome?

(iv) How is the activity of Nedd4-2 itself regulated in the cell?

(v) What is the physiological function of Nedd4 proteins in vivo in mammals, especially in the kidney? The latter should be answered with the generation of knockout murine models for these proteins (not yet published). Future work will undoubtedly address these and other important questions that investigate the relationship between ENaC and the ubiquitin system.

List of Abbreviations Used

CCT: cortical collecting tubule; CNT: distal connecting tubule; ENaC: epithelial Na+ channel; HECT: homologous to E6AP carboxyl-terminus; PHAI: pseudohypoaldosteronism I; PY: Pro-Pro-x-Tyr.

Competing Interests

The author declares that they have no competing interests.

Acknowledgements

Work from the author's lab described in this review was supported by grants from the Canadian Institute of Health Research and the Canadian CF Foundation.

References

1. Burt VL, Whelton P, Roccella EJ, Brown C, Cutler JA, Higgins M, Horan MJ, Labarthe D: Prevalence of hypertension in the US adult population. Results from the Third National Health and Nutrition Examination Survey, 1988–1991. Hypertension 1995, 25(3):305–313.

2. Lifton RP, Gharavi AG, Geller DS: Molecular mechanisms of human hypertension. Cell 2001, 104(4):545–556.

3. Rossier BC: The epithelial sodium channel: activation by membrane-bound serine proteases. Proc Am Thorac Soc 2004, 1(1):4–9.

4. Duc C, Farman N, Canessa CM, Bonvalet JP, Rossier BC: Cell-specific expression of epithelial sodium channel alpha, beta, and gamma subunits in aldosterone-responsive epithelia from the rat: localization by in situ hybridization and immunocytochemistry. J Cell Biol 1994, 127(6 Pt 2):1907–1921.

5. Frindt G, Palmer LG: Na channels in the rat connecting tubule. Am J Physiol Renal Physiol 2004, 286(4):F669–674.

6. Garty H, Palmer LG: Epithelial sodium channels: function, structure, and regulation. Physiol Rev 1997, 77(2):359–396.

7. Kellenberger S, Schild L: Epithelial sodium channel/degenerin family of ion channels: a variety of functions for a shared structure. Physiol Rev 2002, 82(3):735–767.

8. Schild L: The epithelial sodium channel: from molecule to disease. Rev Physiol Biochem Pharmacol 2004, 151:93 107.

9. Canessa CM, Schild L, Buell G, Thorens B, Gautschi I, Horisberger JD, Rossier BC: Amiloride-sensitive epithelial Na+ channel is made of three homologous subunits. Nature 1994, 367(6462):463–467.

10. Firsov D, Gautschi I, Merillat AM, Rossier BC, Schild L: The heterotetrameric architecture of the epithelial sodium channel (ENaC). Embo J 1998, 17(2):344–352.

11. Kosari F, Sheng S, Li J, Mak DO, Foskett JK, Kleyman TR: Subunit stoichiometry of the epithelial sodium channel. J Biol Chem 1998, 273(22):13469–13474.

12. Snyder PM, Cheng C, Prince LS, Rogers JC, Welsh MJ: Electrophysiological and biochemical evidence that DEG/ENaC cation channels are composed of nine subunits. J Biol Chem 1998, 273(2):681–684.

13. Staruschenko A, Medina JL, Patel P, Shapiro MS, Booth RE, Stockand JD: Fluorescence resonance energy transfer analysis of subunit stoichiometry of the epithelial Na+ channel. J Biol Chem 2004, 279(26):27729–27734.

14. Strautnieks SS, Thompson RJ, Gardiner RM, Chung E: A novel splice-site mutation in the gamma subunit of the epithelial sodium channel gene in three pseudohypoaldosteronism type 1 families. Nat Genet 1996, 13(2):248–250.

15. Chang SS, Grunder S, Hanukoglu A, Rosler A, Mathew PM, Hanukoglu I, Schild L, Lu Y, Shimkets RA, Nelson-Williams C, et al.: Mutations in subunits

of the epithelial sodium channel cause salt wasting with hyperkalaemic acidosis, pseudohypoaldosteronism type 1. Nat Genet 1996, 12(3):248–253.

16. Barker PM, Nguyen MS, Gatzy JT, Grubb B, Norman H, Hummler E, Rossier B, Boucher RC, Koller B: Role of gammaENaC subunit in lung liquid clearance and electrolyte balance in newborn mice. Insights into perinatal adaptation and pseudohypoaldosteronism. J Clin Invest 1998, 102(8):1634–1640.

17. McDonald FJ, Yang B, Hrstka RF, Drummond HA, Tarr DE, McCray PB, Stokes JB, Welsh MJ, Williamson RA: Disruption of the beta subunit of the epithelial Na+ channel in mice: hyperkalemia and neonatal death associated with a pseudohypoaldosteronism phenotype. Proc Natl Acad Sci USA 1999, 96(4):1727–1731.

18. Hummler E, Barker P, Talbot C, Wang Q, Verdumo C, Grubb B, Gatzy J, Burnier M, Horisberger JD, Beermann F, et al.: A mouse model for the renal salt-wasting syndrome pseudohypoaldosteronism. Proc Natl Acad Sci USA 1997, 94(21):11710–11715.

19. Grunder S, Firsov D, Chang SS, Jaeger NF, Gautschi I, Schild L, Lifton RP, Rossier BC: A mutation causing pseudohypoaldosteronism type 1 identifies a conserved glycine that is involved in the gating of the epithelial sodium channel. Embo J 1997, 16(5):899–907.

20. Liddle GW, Bledsoe T, Coppage WS Jr: Hypertension reviews. J Tenn Med Assoc 1974, 67(8):669.

21. Snyder PM, Price MP, McDonald FJ, Adams CM, Volk KA, Zeiher BG, Stokes JB, Welsh MJ: Mechanism by which Liddle's syndrome mutations increase activity of a human epithelial Na+ channel. Cell 1995, 83(6):969–978.

22. Firsov D, Schild L, Gautschi I, Merillat AM, Schneeberger E, Rossier BC: Cell surface expression of the epithelial Na channel and a mutant causing Liddle syndrome: a quantitative approach. Proc Natl Acad Sci USA 1996, 93(26):15370–15375.

23. Shimkets RA, Warnock DG, Bositis CM, Nelson-Williams C, Hansson JH, Schambelan M, Gill JR Jr, Ulick S, Milora RV, Findling JW, et al.: Liddle's syndrome: heritable human hypertension caused by mutations in the beta subunit of the epithelial sodium channel. Cell 1994, 79(3):407–414.

24. Hansson JH, Nelson-Williams C, Suzuki H, Schild L, Shimkets R, Lu Y, Canessa C, Iwasaki T, Rossier B, Lifton RP: Hypertension caused by a truncated epithelial sodium channel gamma subunit: genetic heterogeneity of Liddle syndrome. Nat Genet 1995, 11(1):76–82.

25. Hansson JH, Schild L, Lu Y, Wilson TA, Gautschi I, Shimkets R, Nelson-Williams C, Rossier BC, Lifton RP: A de novo missense mutation of the beta

subunit of the epithelial sodium channel causes hypertension and Liddle syndrome, identifying a proline-rich segment critical for regulation of channel activity. Proc Natl Acad Sci USA 1995, 92(25):11495–11499.

26. Schild L, Lu Y, Gautschi I, Schneeberger E, Lifton RP, Rossier BC: Identification of a PY motif in the epithelial Na channel subunits as a target sequence for mutations causing channel activation found in Liddle syndrome. Embo J 1996, 15(10):2381–2387.

27. Tamura H, Schild L, Enomoto N, Matsui N, Marumo F, Rossier BC: Liddle disease caused by a missense mutation of beta subunit of the epithelial sodium channel gene. J Clin Invest 1996, 97(7):1780–1784.

28. Inoue J, Iwaoka T, Tokunaga H, Takamune K, Naomi S, Araki M, Takahama K, Yamaguchi K, Tomita K: A family with Liddle's syndrome caused by a new missense mutation in the beta subunit of the epithelial sodium channel. J Clin Endocrinol Metab 1998, 83(6):2210–2213.

29. Kanelis V, Rotin D, Forman-Kay JD: Solution structure of a Nedd4 WW domain – ENaC peptide complex. Nature Structure Biol 2001, 8(5):407–412.

30. Staub O, Dho S, Henry P, Correa J, Ishikawa T, McGlade J, Rotin D: WW domains of Nedd4 bind to the proline-rich PY motifs in the epithelial Na+ channel deleted in Liddle's syndrome. Embo J 1996, 15(10):2371–2380.

31. Hershko A, Ciechanover A: The ubiquitin system. Annu Rev Biochem 1998, 67:425–479.

32. Glickman MH, Ciechanover A: The ubiquitin-proteasome proteolytic pathway: destruction for the sake of construction. Physiol Rev 2002, 82(2):373–428.

33. Hicke L: A new ticket for entry into budding vesicles-ubiquitin. Cell 2001, 106(5):527–530.

34. Staub O, Rotin D: Role of ubiquitylation in cellular membrane transport. Physiol Rev 2006, 86(2):669–707.

35. Hicke L, Schubert HL, Hill CP: Ubiquitin-binding domains. Nat Rev Mol Cell Biol 2005, 6(8):610–621.

36. Plant PJ, Lafont F, Lecat S, Verkade P, Simons K, Rotin D: Apical membrane targeting of Nedd4 is mediated by an association of its C2 domain with annexin XIIIb. J Cell Biol 2000, 149(7):1473–1484.

37. Plant PJ, Yeger H, Staub O, Howard P, Rotin D: The C2 domain of the ubiquitin protein ligase Nedd4 mediates Ca2+-dependent plasma membrane localization. J Biol Chem 1997, 272(51):32329–32336.

38. Goulet CC, Volk KA, Adams CM, Prince LS, Stokes JB, Snyder PM: Inhibition of the epithelial Na+ channel by interaction of Nedd4 with a PY motif deleted in Liddle's syndrome. J Biol Chem 1998, 273(45):30012–30017.

39. Harvey KF, Dinudom A, Komwatana P, Jolliffe CN, Day ML, Parasivam G, Cook DI, Kumar S: All three WW domains of murine Nedd4 are involved in the regulation of epithelial sodium channels by intracellular Na+. J Biol Chem 1999, 274(18):12525–12530.

40. Farr TJ, Coddington-Lawson SJ, Snyder PM, McDonald FJ: Human Nedd4 interacts with the human epithelial Na+ channel: WW3 but not WW1 binds to Na+-channel subunits. Biochem J 2000, 345(Pt 3):503–509.

41. Henry PC, Kanelis V, O'Brien MC, Kim B, Gautschi I, Forman-Kay J, Schild L, Rotin D: Affinity and specificity of interactions between Nedd4 isoforms and the epithelial Na+ channel. J Biol Chem 2003, 278(22):20019–20028.

42. Kanelis V, Bruce MC, Skrynnikov NR, Rotin D, Forman-Kay JD: Structural determinants for high-affinity binding in a Nedd4 WW3* domain-Comm PY motif complex. Structure 2006, 14(3):543–553.

43. Rotin D, Staub O, Haguenauer-Tsapis R: Ubiquitination and endocytosis of plasma membrane proteins: role of Nedd4/Rsp5p family of ubiquitin-protein ligases. J Membr Biol 2000, 176(1):1–17.

44. Ingham RJ, Gish G, Pawson T: The Nedd4 family of E3 ubiquitin ligases: functional diversity within a common modular architecture. Oncogene 2004, 23(11):1972–1984.

45. Abriel H, Loffing J, Rebhun JF, Pratt JH, Schild L, Horisberger JD, Rotin D, Staub O: Defective regulation of the epithelial Na+ channel by Nedd4 in Liddle's syndrome. J Clin Invest 1999, 103(5):667–673.

46. Kamynina E, Debonneville C, Bens M, Vandewalle A, Staub O: A novel mouse Nedd4 protein suppresses the activity of the epithelial Na+ channel. Faseb J 2001, 15(1):204–214.

47. Snyder PM, Olson DR, McDonald FJ, Bucher DB: Multiple WW domains, but not the C2 domain, are required for inhibition of the epithelial Na+ channel by human Nedd4. J Biol Chem 2001, 276(30):28321–28326.

48. Staub O, Gautschi I, Ishikawa T, Breitschopf K, Ciechanover A, Schild L, Rotin D: Regulation of stability and function of the epithelial Na+ channel (ENaC) by ubiquitination. Embo J 1997, 16(21):6325–6336.

49. Lu C, Pribanic S, Debonneville A, Jiang C, Rotin D: The PY motif of ENaC, mutated in Liddle syndrome, regulates channel internalization, sorting and mobilization from subapical pool. Traffic 2007, in press.

50. Kamynina E, Tauxe C, Staub O: Distinct characteristics of two human Nedd4 proteins with respect to epithelial Na(+) channel regulation. Am J Physiol Renal Physiol 2001, 281(3):F469–477.

51. Shimkets RA, Lifton RP, Canessa CM: The activity of the epithelial sodium channel is regulated by clathrin-mediated endocytosis. J Biol Chem 1997, 272(41):25537–25541.

52. Pradervand S, Wang Q, Burnier M, Beermann F, Horisberger JD, Hummler E, Rossier BC: A mouse model for Liddle's syndrome. J Am Soc Nephrol 1999, 10(12):2527–2533.

53. Palmer LG, Sackin H, Frindt G: Regulation of Na+ channels by luminal Na+ in rat cortical collecting tubule. J Physiol 1998, 509(Pt 1):151–162.

54. Kellenberger S, Gautschi I, Rossier BC, Schild L: Mutations causing Liddle syndrome reduce sodium-dependent downregulation of the epithelial sodium channel in the Xenopus oocyte expression system. J Clin Invest 1998, 101(12):2741–2750.

55. Chen SY, Bhargava A, Mastroberardino L, Meijer OC, Wang J, Buse P, Firestone GL, Verrey F, Pearce D: Epithelial sodium channel regulated by aldosterone-induced protein sgk. Proc Natl Acad Sci USA 1999, 96(5):2514–2519.

56. Naray-Fejes-Toth A, Canessa C, Cleaveland ES, Aldrich G, Fejes-Toth G: sgk is an aldosterone-induced kinase in the renal collecting duct. Effects on epithelial na+ channels. J Biol Chem 1999, 274(24):16973–16978.

57. Debonneville C, Flores SY, Kamynina E, Plant PJ, Tauxe C, Thomas MA, Munster C, Chraibi A, Pratt JH, Horisberger JD, et al.: Phosphorylation of Nedd4-2 by Sgk1 regulates epithelial Na(+) channel cell surface expression. Embo J 2001, 20(24):7052–7059.

58. Snyder PM, Olson DR, Thomas BC: Serum and glucocorticoid-regulated kinase modulates Nedd4-2-mediated inhibition of the epithelial Na+ channel. J Biol Chem 2002, 277(1):5–8.

59. Kjarland E, Keen TJ, Kleppe R: Does isoform diversity explain functional differences in the 14–3–3 protein family? Curr Pharm Biotechnol 2006, 7(3):217–223.

60. Ichimura T, Yamamura H, Sasamoto K, Tominaga Y, Taoka M, Kakiuchi K, Shinkawa T, Takahashi N, Shimada S, Isobe T: 14–3–3 proteins modulate the expression of epithelial Na+ channels by phosphorylation-dependent interaction with Nedd4-2 ubiquitin ligase. J Biol Chem 2005, 280(13):13187–13194.

61. Bhalla V, Daidie D, Li H, Pao AC, LaGrange LP, Wang J, Vandewalle A, Stockand JD, Staub O, Pearce D: Serum- and glucocorticoid-regulated kinase 1 regulates

ubiquitin ligase neural precursor cell-expressed, developmentally down-regulated protein 4-2 by inducing interaction with 14–3–3. Mol Endocrinol 2005, 19(12):3073–3084.

62. Liang X, Peters KW, Butterworth MB, Frizzell RA: 14–3–3 isoforms are induced by aldosterone and participate in its regulation of epithelial sodium channels. J Biol Chem 2006, 281(24):16323–16332.

63. Flores SY, Loffing-Cueni D, Kamynina E, Daidie D, Gerbex C, Chabanel S, Dudler J, Loffing J, Staub O: Aldosterone-induced serum and glucocorticoid-induced kinase 1 expression is accompanied by Nedd4-2 phosphorylation and increased Na+ transport in cortical collecting duct cells. J Am Soc Nephrol 2005, 16(8):2279–2287.

64. Loffing-Cueni D, Flores SY, Sauter D, Daidie D, Siegrist N, Meneton P, Staub O, Loffing J: Dietary sodium intake regulates the ubiquitin-protein ligase nedd4-2 in the renal collecting system. J Am Soc Nephrol 2006, 17(5):1264–1274.

65. Alvarez de la Rosa D, Zhang P, Naray-Fejes-Toth A, Fejes-Toth G, Canessa CM: The serum and glucocorticoid kinase sgk increases the abundance of epithelial sodium channels in the plasma membrane of Xenopus oocytes. J Biol Chem 1999, 274(53):37834–37839.

66. Shigaev A, Asher C, Latter H, Garty H, Reuveny E: Regulation of sgk by aldosterone and its effects on the epithelial Na(+) channel. Am J Physiol Renal Physiol 2000, 278(4):F613–619.

67. Auberson M, Hoffmann-Pochon N, Vandewalle A, Kellenberger S, Schild L: Epithelial Na+ channel mutants causing Liddle's syndrome retain ability to respond to aldosterone and vasopressin. Am J Physiol Renal Physiol 2003, 285(3):F459–471.

68. Diakov A, Korbmacher C: A novel pathway of epithelial sodium channel activation involves a serum- and glucocorticoid-inducible kinase consensus motif in the C terminus of the channel's alpha-subunit. J Biol Chem 2004, 279(37):38134–38142.

69. Dahlmann A, Pradervand S, Hummler E, Rossier BC, Frindt G, Palmer LG: Mineralocorticoid regulation of epithelial Na+ channels is maintained in a mouse model of Liddle's syndrome. Am J Physiol Renal Physiol 2003, 285(2):F310–318.

70. Morris RG, Schafer JA: cAMP Increases Density of ENaC Subunits in the Apical Membrane of MDCK Cells in Direct Proportion to Amiloride-sensitive Na(+) Transport. J Gen Physiol 2002, 120(1):71–85.

71. Butterworth MB, Edinger RS, Johnson JP, Frizzell RA: Acute ENaC stimulation by cAMP in a kidney cell line is mediated by exocytic insertion from a recycling channel pool. J Gen Physiol 2005, 125(1):81–101.

72. Snyder PM: Liddle's syndrome mutations disrupt cAMP-mediated translocation of the epithelial Na(+) channel to the cell surface. J Clin Invest 2000, 105(1):45–53.

73. Snyder PM, Olson DR, Kabra R, Zhou R, Steines JC: cAMP and serum and glucocorticoid-inducible kinase (SGK) regulate the epithelial Na(+) channel through convergent phosphorylation of Nedd4-2. J Biol Chem 2004, 279(44):45753–45758.

74. Vallet V, Chraibi A, Gaeggeler HP, Horisberger JD, Rossier BC: An epithelial serine protease activates the amiloride-sensitive sodium channel. Nature 1997, 389(6651):607–610.

75. Hughey RP, Bruns JB, Kinlough CL, Harkleroad KL, Tong Q, Carattino MD, Johnson JP, Stockand JD, Kleyman TR: Epithelial sodium channels are activated by furin-dependent proteolysis. J Biol Chem 2004, 279(18):18111-18114.

76. Hughey RP, Bruns JB, Kinlough CL, Kleyman TR: Distinct pools of epithelial sodium channels are expressed at the plasma membrane. J Biol Chem 2004, 279(47):48491–48494.

77. Hughey RP, Mueller GM, Bruns JB, Kinlough CL, Poland PA, Harkleroad KL, Carattino MD, Kleyman TR: Maturation of the epithelial Na+ channel involves proteolytic processing of the alpha- and gamma-subunits. J Biol Chem 2003, 278(39):37073–37082.

78. Knight KK, Olson DR, Zhou R, Snyder PM: Liddle's syndrome mutations increase Na+ transport through dual effects on epithelial Na+ channel surface expression and proteolytic cleavage. Proc Natl Acad Sci USA 2006, 103(8):2805–2808.

A Novel Mechanism in Regulating the Alpha-Subunit of the Epithelial Sodium Channel (α ENaC) bythe Alternatively Spliced Form α ENaC-b

Marlene F. Shehata

ABSTRACT

Introduction

In Dahl rats' kidney cortex, the alternatively spliced form of the epithelial sodium channel α subunit (α ENaC-b) is the most abundant mRNA transcript (32+/-3 fold > α ENaC-wt) as was investigated by quantitative

RT-PCR analysis. α ENaC-b mRNA levels were significantly higher in Dahl R versus S rats, and were further augmented by high salt diet.

Objectives

In the present study, we described the molecular cloning and searched for a possible role of α ENaC-b by testing its potential expression in COS7 cells as well as its impact on α ENaC-wt expression levels when co-expressed in COS7 cells in a dose-dependent manner.

Methods

Using RT-PCR strategy, the full-length wildtype α ENaC transcript and the alternatively spliced form α ENaC-b were amplified, sequenced, cloned, sub-cloned into PCMV-sport6 expression vector, expressed and co-expressed into COS7 cells in a dose-dependent manner. A combination of denaturing and native western blotting techniques was employed to examine the expression of α ENaC-b in vitro, and to determine if an interaction between α ENaC-b and α ENaC-wt occurs in vitro, and finally to demonstrate if degradation of α ENaC-wt protein does occur.

Results

α ENaC-b is translated in COS7 cells. Co-expression of α ENaC-b together with α ENaC-wt reduced α ENaC-wt levels in a dose-dependent manner. α ENaC-wt and α ENaC-b appear to form a complex that enhances the degradation of α ENaC-wt.

Conclusions: Western blots suggest a novel mechanism in α ENaC regulation whereby α ENaC-b exerts a dominant negative effect on α ENaC-wt expression. This is potentially by sequestering α ENaC-wt, enhancing its proteolytic degradation, and possibly explaining the mechanism of salt-resistance in Dahl R rats.

Keywords: *α ENaC-wt, α ENaC-b, COS7 cells, transfection, expression, binding, dose-dependent response, salt-sensitive hypertension, Dahl rats*

Introduction

The amiloride-sensitive epithelial sodium channels (ENaC) consists of at least three subunits denoted by α, β, and γ, each of which possesses two transmembrane domains.[1-3] The α ENaC subunit is a key molecule for the formation of a functional Na^+ channel complex in vivo, whereas the β and γ subunits greatly potentiate the channel activity.[3] α ENaC gene knockout in mice was lethal within

40 h of birth because of fluid-filled lungs, confirming the critical role of α ENaC in forming functional Na⁺ channels in vivo.[4] Naturally occurring α ENaC alternatively spliced forms have been identified in mice, chicken, humans and rats.[5-8] To date, there are two alternatively spliced forms of the α subunit of ENaC (α ENaC-a and -b) that have been published in rats, in addition to the major α ENaC transcript.[7] α ENaC-a transcript is a low abundance transcript compared to the full-length α ENaC and has been studied in terms of expression, functionality and binding to ENaC blocker.[7] On the other hand, α ENaC-b is yet to be characterized and appears to play a role in modulating ENaC for the following reasons: a) it shares the same splice site with human and chicken spliced forms; b) our preliminary results showed that unlike α ENaC-a, α ENaC-b, mRNA levels are □32 ± 3-fold higher than full-length α ENaC in kidneys of Dahl rats indicating an increased stability/synthesis of the former;[9] c) α ENaC-b mRNA concentrations are significantly higher in Dahl R versus Dahl S rats,[9] suggesting a putative dominant negative effect on α ENaC activity in a model of suppressed ENaC activity (Dahl R rats);[10] d) α ENaC-b is a salt-sensitive transcript, because high versus normal salt diet caused a remarkable increase in α ENaC-b mRNA levels ($P < 0.05$) in Dahl R rats.[9]

Owing to the fact that non functional α ENaC alternatively spliced forms, such as α ENaC-b, have been proposed to serve as negative regulatory components for ENaC activity,8 we were prompted to invest in basic research with a view to identifying novel targets for diagnosis, prevention and treatment of disorders related to ENaC dysfunction. The overall objectives of these studies were to investigate whether α ENaC-b is translated in vitro, whether it is co-expressed with α ENaC-wt, and whether α ENaC-b expression suppresses α ENaC-wt expression when co-expressed in COS7 cells in a dose-dependent manner.

Methods

RNA Extraction, RT-PCR and PCR

Kidney cortex of Dahl S rats was cut out micro¬scopically and homogenized with polytron. Poly (A)⁺ mRNA from kidney cortex of Dahl S rats was isolated using Trizol reagent (Invitrogen, CA, U. S.A.) according to the manufacturer's protocol. Isolated RNA was subjected to DNAse treatment for removing potential genomic DNA contamination using DNase treatment kit (Ambion, ON, Canada). First strand cDNA was synthesized from 2 μg of mRNA with superscript II RNase H- Reverse transcriptase (Invitrogen, CA, U.S.A.).

For α ENaC-wt and α ENaC-b, specifi c primers (Table 1) were designed to flank the open reading frames of α ENaC-wt and -b form. High fidelity Expand

Long Range, dNTPack (kind gift from Roche Applied Science®, QC, Canada) is employed to amplify the full length α ENaC and α ENaC-b. Touchdown PCR method was employed to improve the efficiency of gene amplification as follows (start temperature (Tm) at 68 °C and the annealing Tm reduced 2 °C per cycle for the next cycles up to Tm of 58 °C, then the remaining cycles were performed at a 58 °C annealing temperature). After amplification, the specific fragments of 2267 bp (α ENaC-wt) and 1586 bp (α ENaC-b) were visualized in 1% agarose gels (Fig. 1).

	Primers used	Amplicon size
α ENaC-wt	**Sense:** 5′-AGCTCAATACTGCTTGGTTGG-3′ **Antisense:** 5′-AGGGCTGGGTGAGAGGAT-3′	2.2 kb
α ENaC-b	**Sense:** 5′- AGCTCAATACTGCTTGGTTGG-3′ **Antisense:** 5′- AAAGCAGAGCTCCTGGGTGT-3′	1.6 kb

Sense and antisense primers were selected to amplify full length α ENaC-wt and -b transcripts. The expected amplicon size was 2.2 kb for α ENaC-wt and 1.6 kb for α ENaC-b.

Table 1. Primers used for reverse transcription polymerase chain reaction and expected sizes of cDNA fragments.

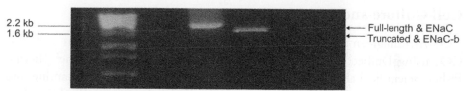

Figure 1. Amplification of ENaC-wt and alternatively spliced form from total RNA. Total RNA extracted from kidney cortex of Dahl S rats was utilized as a template to generate full lengths cDNAs of α ENaC-wt and alternatively spliced form α ENaC-b. High fidelity Expand Long Range, dNTPack was used to amplify α ENaC-wt and -b form. Touchdown PCR method was employed to improve the effi ciency of gene amplifi cation. After amplifi cation, the specifi c fragments of 2267 bp (α ENaC-wt) and 1586 bp (α ENaC-b) were visualized in 1% agarose gels.

Plasmids and Generation of Constructs

PCR products for α ENaC-wt and α ENaC-b were ligated with pCR˚2.1-TOPO˚ vector using the TOPO TA˚ cloning kit (Invitrogen, CA, U.S.A.), with blue/white screening of DH5α competent cells on ampicillin selective plates. A total of 14 individual colonies for each species were isolated. Positive clones were picked and cultured in LB broth. The DNA extracted from these cultures using Qiagen DNA miniprep kit, was sequenced using the M13 universal forward and reverse primers, in addition to embedded primers within α ENaC-wt and-b forms to verify their full lengths sequences. Sequencing was performed using the DYEnamic ET

Terminator kit according to the instructions provided by the manufacturer (PE Applied Biosystems, Foster City, CA). Sequencing products were purified (DyeEx 2.0 spin kit columns; Qiagen Canada, Mississauga, ON, Canada) and analyzed on 3100 DNA analyser (Applied Biosystems, Foster City, CA). The sequence of the cloned α ENaC-b is identical to the one previously reported,[7] and the sequence of the α ENaC-wt is identical to the one previously reported by our group for Dahl rats.[11]

Ligated pCR®2.1-TOPO®-α ENaC-wt clone was digested using EcoRI, while that of α ENaC-b was digested using EcoRV and HindIII. Digested α ENaC-wt and -b fragments were purifi ed from 1% agarose gel (QIAquick Gel Extraction Kit, QiagenTM, ON, Canada) and then subcloned overnight at 16 °C into PC-MV-sport6 vector (Invitrogen, CA, U.S.A.) downstream the T7 promotor using the EcoRI site for α ENaC-wt and the EcoRV and HindIII sites for α ENaC-b. The positive clones were verified by restriction digestion, then PCR of the inserted construct and later by direct nucleotide sequencing as described above. DNA yield from positive clones was later increased for transfection studies using Qiagen midi-prep kit (QiagenTM, ON, Canada) for amplification of the yield.

Cell Culture and Transfection

COS-African green monkey kidney cells were maintained in culture at 37°C/5% CO_2 using Dulbecco's Modified Eagle's Medium (DMEM, Hyclone® Thermo Fisher Scientific Laboratories, Logan, UT) supplemented with glutamine and 10% heat-inactivated fetal bovine serum (FBS, PAA Laboratories, Pasching, Austria). Transfections were performed using lipofectamine 2000 reagent (Invitrogen, Carlsbad, CA, U.S.A.) and up to 35 µg DNA per 10 cm plate in co-transfection experiments. COS7 cells were seeded at approximately one third confluency on 10 cm diameter plates 18 h before the transfection. Immediately before transfection, the culture medium was replaced with 5 ml of DMEM supplemented with 10% (v/v) fetal bovine serum. PCMV-sport6 /αENaC-wt and -b were transfected into COS7 cells separately at a concentration of 25 µg DNA /plate (Fig. 2), as well as co-transfected together in a dose-dependent manner as demonstrated in Figures 3 and 4. The empty PCMV-sport6 vector at a concentration of 25 µg DNA /plate was transfected as a control. DNA was added to 250 µl TE buffer and lipofectamine was then added (Invitrogen®, ON, Canada). The mixture was incubated at room temperature for 20 min before addition to the COS7 cell culture. The cells were incubated overnight at 37 °C in an air/CO2 (19:1) atmosphere before the medium was aspirated and the COS7 cell culture osmotically shocked for 2 min with 10% (v/v) PBS.

Electrophoresis Analysis

We used two different Western blot analyses namely SDS-PAGE or "denaturing gels" and native PAGE or "non denaturing gels". While SDS-PAGE denatured α ENaC-wt and -b proteins, and allowed an easy access of the α ENaC antibody to the reactive epitope and subsequent separation of the α ENaC proteins based on molecular mass, native gel electrophoresis, on the other hand, allowed us to study the α ENaC-wt and -b proteins oligomerization potential. Migration for both SDS-PAGE and native gels was performed under 150 V and for 120 min.

SDS-PAGE and Western Blot Analyses

Cells from transiently transfected COS7 cells expressing α ENaC-wt, or α ENaC-b, or the empty vector along with COS7 cells equally co-transfected with α ENaC-wt and -b form were washed twice with ice-cold PBS, scraped and then maintained Shehata et al for >2 h at 4 °C in RIPA lysis buffer (10 mM $NaPO_4$, 150 mM NaCl, 1% deoxycholate, 1% Triton X-100, and 0.1% SDS, pH 7.2) supplemented with the protease inhibitor cocktail (Sigma, St. Louis, MO). Protein concentrations were measured by Bradford. Proteins were then separated by SDS-PAGE resolving gels containing 10% acrylamide, and 4% of the same solution for stacking gels. Samples were mixed with Laemmli buffer (Bio-Rad, 0.5 M Tris-Hcl, pH 6.8, 10% glycerol, 0.02% bromophenol blue) containing 0.1% SDS [sodium dodecyl-sulphate (w:v)] and supplemented with β-mercaptoethanol (v:v). Before loading, SDS samples were heated at 100°C for 5 minutes. Proteins were separated by SDS-PAGE and western blots were performed according to procedures recommended by the Bio-Rad Protein III system (Bio-Rad Inc.).

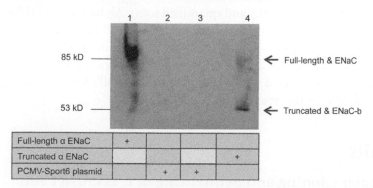

Figure 2. Expression of α EN αC-wt and α ENαC-b in COS7 cells. SDS/PAGE and Western blot analysis were performed using extracts from mammalian cells expressing either the alternatively spliced form α ENaC-wt or –b form or the empty vector in COS7 cells. Lane 1 represents proteins extrated from COS7 cells transfected with α ENaC-wt, lanes 2 and 3 represent COS7 cells transfected with the empty PCMV-sport6 vector, and lane 4 represents COS7 cells transfected with α ENaC-b.

Native Gel Electrophoresis and Western Blot Analyses

Separation of proteins by PAGE under non denaturing conditions was performed similar to SDS-PAGE, except that SDS and β-mercaptoethanol were totally excluded and the samples were not boiled. This will preserve any protein interactions. Samples and protein markers were diluted in 0.5 M Tris-HCl, pH 6.8, 40% glycerol (v:v). and 0.02% bromophenol blue (w:v) (without SDS and β-mercaptoethanol), and without heating. In Native PAGE, 20 μg of proteins were separated on precast Novex° 8% Tris-Glycine Midi Gel (Invitrogen, CA, U.S.A.). Before loading the samples, the gel was cooled to 4 °C. After addition of native sample buffer (5X), samples were directly applied to the gel without any preconditioning. Native gel electrophoresis (18 cm • 16 cm) was conducted in the same apparatus as above. The running buffer (pH 8.3) contained Tris base (25 mM) and glycine (192 mM).

Polyclonal antibodies raised against the N terminal aa 46–68 [NH2-LGKGD-KREEQGL-GPEPSAPRQPT-COOH] of the rat α ENaC were used to recognize the α ENaC-wt and -b proteins. Antibodies were purified from rabbit sera, concentrated to 1.3 ug/ul and diluted to 1:1000. Primary α ENaC antibodies were a kind gift from Dr. M. Amin. In Western blot, protein samples were transferred to nitrocellulose membranes (Bio-Rad Inc.) using Trans-Blot Cell system (Bio-Rad Inc., CA, U.S.A.). After transfer, nitrocellulose membranes were blocked for 1 h with 5% skimmed milk in 0.1% Tris-buffered saline containing 0.1% Triton X-100 (TBST) and incubated with primary antibody overnight. Blots were probed with horseradish peroxidase-conjugated anti-rabbit IgG secondary antibody (Promega Inc., ON, Canada) for 2 h at room temperature. Multiple washes for the membranes were performed using TBST after the primary and secondary incubations. Protein signals were detected using ECL substrate (Amersham Biosciences Inc. Chicago, IL) and quantified using Alpha Ease FC (FluorChem, 9900) software. All Western blots were stripped in 100 mm 2-mercaptoethanol, 62.5 mM Tris-HCl (pH 6.7), and 2% SDS for 30 min at 55 °C with constant agitation. After removal of antibodies, nonspecific interactions were reblocked by incubation in Tris-buffered saline-Tween 20 and 5% milk for 2 h before the blots were reprobed with the house keeping gene (β-actin).

Results

Molecular Cloning and Sequencing of α ENaC-wt and Alternatively Spliced Form a ENaC-b

To determine whether α ENaC-b is translated in vitro, and if so, whether α ENaC-b binds to α ENaC-wt, we cloned α ENaC-wt and -b transcripts from Dahl

S rats' kidney cortex, using RT-PCR. Using PCR primers that were designed to amplify full-length coding sequences of α ENaC-wt and α ENaC-b transcripts, α ENaC-wt and α ENaC-b were obtained and cloned (Fig. 1). Sequences for both α ENaC-wt and -b form were confirmed by complete nucleotide sequencing analyses. α ENaC-wt sequence was consistent with our previously reported Dahl rat coding sequence.11 The present α ENaC-b sequence was consistent with the previously reported α ENaC-b sequence.7

Expression of Full Length a ENaC and Truncated α ENaC-b Alternatively spliced Form in Transiently Transfected COS7 Cells

Full length α ENaC and truncated α ENaC-b transcripts were studied in COS7 cells to confirm their ability to make a protein. In vitro translation of α ENaC-wt and α ENaC-b produced proteins of 85 and 53 kDa, respectively (Fig. 2).

Dimerization of α ENaC-wt and α ENaC-b

Using native Western analysis for lysates obtained from COS7 cells co-transfected with α ENaC-wt and -b forms, a dimer was formed at 140 kDa, equivalent to the summation of the molecular weights of α ENaC-wt and α ENaC-b. α ENaC-wt protein has a molecular mass of 85 kDa, while α ENaC-b protein has a molecular mass of 53 kDa (Fig. 3).

Dose-Dependent Suppression of Full-Length α ENaC Expressionby Increasing Doses of α ENaC-b

Western blot analysis using an α ENaC antiserum revealed that co-expression of α ENaC-wt with spliced form α ENaC-b decreased α ENaC-wt protein levels by increasing doses of α ENaC-b protein (Fig. 4).

Discussion

The present study reports the following: a) α ENaC-b transcript is translated in COS7 cells to a peptide of apparent mass of 53 kDa, b) α ENaC-b transcript is co-expressed with α ENaC-wt transcript, c) α ENaC-b forms a complex with α ENaC-wt resulting in dimer formation of a molecular mass of approximately 140 kDa, d) Increasing doses of α ENaC-b suppressed α ENaC-wt expression in a dose-dependent manner when co-expressed in COS7 cells.

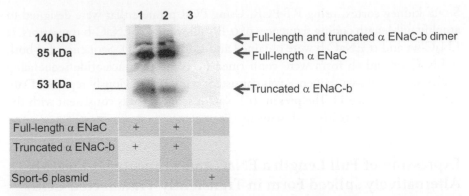

Figure 3. Analysis of α ENaC-wt and α ENaC-b expression by native Western analysis. α ENaC-wt and α ENaC-b interacts in vivo by the formation of a dimmer of a molecular mass of 140 kDa. Lanes 1 and 2 represent proteins from COS7 cells co-expressing α ENaC-wt and α ENaC-b in equal proportions (1:1), lane 3 represents proteins from COS7 cells transfected with the empty PCMV-sport6 vector. Bands are normalized against the house keeping gene ® -actin and the experiment was done in triplicate.

Formation of alternatively spliced forms of α ENaC have been previously reported in four species: rats, humans, mice and chicken6–9,12 suggesting that alternative RNA splicing is possibly a mechanism for regulating α ENaC activity, besides transcriptional regulation. The biological impact of alternatively spliced forms, particularly those lacking functional domains such as α ENaC-b lacking the second transmembrane domain, may go as far as a drastic switch-off effect.13–15

Our previous findings showed that α ENaC-b mRNA levels were significantly higher in Dahl salt-resistant (R) versus salt-sensitive (S) rats.9 Overrepresentation of α ENaC-b in Dahl R rats could be indicative of a dominant negative effect on α ENaC expression/function in this model. Additionally, owing to the fact that α ENaC-b is a salt-sensitive transcript in Dahl R rats,9 we were motivated to search for a biological role of α ENaC-b in vitro that might be in progress to protect the channel from over-activity in high salt environment in Dahl R rats, possibly by a dominant negative effect on α ENaC-wt. Therefore, α ENaC-b was amplified, sequenced and cloned. Expression of α ENaC-b alone and in the presence and absence of α ENaC-wt was examined.

Results show that the truncated α ENaC-b protein is translated in vitro and its co-expression with α ENaC-wt appeared to substantially reduce α ENaC-wt expression in a dose-dependent manner. α ENaC-b acts as a dominant inhibitor of α ENaC-wt by forming a complex with α ENaC-wt. These results support an important role for alternative splicing in α ENaC regulation, and suggest that α ENaC can form heteromers which may be important in regulating α ENaC expression and/or activity.

The present study is the first to demonstrate the expression of alternatively spliced α ENaC-b form in mammalian COS7 cells and expands our understanding of the putative mechanism of action of α ENaC-b as a negative expression regulator of α ENaC-wt; by directly interacting with α ENaC-wt as a dominant negative protein that ultimately hinders α ENaC-wt expression and potentially its activity. In co-expression studies (Fig. 4), a weak signal of α ENaC-b can be detected at 53 kDa, possibly indicating that dimerization of α ENaC-wt with -b protein degrades α ENaC-wt primarily and eventually degrades α ENaC-b. Our findings therefore uncover a new mechanism that may control the regulation of α ENaC expression in vivo by a potential spliced form α ENaC-b. A detailed understanding of the spectrum of alternatively spliced α ENaC-b is of importance not only for the understanding of ENaC regulation in Dahl rats, but also for the future drug development efforts. It will also provide insight into the beneficial impact of increasing α ENaC-b to possibly prevent salt sensitivity.

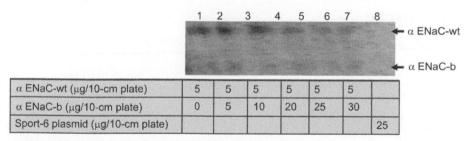

	1	2	3	4	5	6	7	8
α ENaC-wt (μg/10-cm plate)	5	5	5	5	5	5		
α ENaC-b (μg/10-cm plate)	0	5	10	20	25	30		
Sport-6 plasmid (μg/10-cm plate)								25

Figure 4. Dose-dependent suppression of α ENaC-wt by increasing doses of α ENaC-b. Lanes 1–7 represent proteins of COS7 cells co-transfected by α ENaC-wt and α ENaC-b in the ratios provided. Denaturing Western analysis demonstrate a dose-dependent reduction in full-length α ENaC expression. Lane 8 represents proteins from cells transfected with the empty PCMV-sport 6 vector. A weak signal of α ENaC-b can be detected at 53 kDa, possibly indicating that dimerization of α ENaC-wt with α ENaC-b proteins degrades α ENaC-wt primarily and eventually degrades α ENaC-b. Bands are normalized against the house keeping gene β-actin and the experiment was done in triplicate.

To this end, the present study suggests a dominant negative effect imposed by α ENaC-b on full length α ENaC protein expression in Dahl R rat kidneys. Our preliminary results suggest the intriguing possibility of a "genotoxic" effect of α ENaC-b on full-length α ENaC, potentially by accelerating proteolytic degradation of the latter in a dose-dependent fashion. There are at least two areas of potential clinical importance for these studies. First, identification of spliced forms that impair the hyperactivity of ENaC to sodium in the kidney of Dahl S rats would identify specific targets for the development of novel antihypertensive drug therapy for salt-dependent hypertension. Second, as gene therapy delivery systems continue to evolve, we will be examining in vivo delivery of adenoviruses carrying α ENaC-b and then assessing the resultant blood pressure effects. This may eventually gain relevance as a possible "proof-of-concept" gene therapy.

Acknowledgements

The author acknowledges Roche Diagnostics® Canada for generously providing free samples of Taq polymerase enzyme. Additionally, the author is grate¬ful to Qiagen® Canada for providing free samples of the DyeEx 2.0 spin kit columns used for sequencing. MF Shehata is the recipient of the Memorial University of Newfoundland Horizon Award for 2007. MF Shehata has received numerous awards including the Canadian Institutes of Health Research/Pfizer/ Canadian Hypertension Society Doctoral Research Award, Ontario Graduate Scholarship and the Ontario Graduate Scholarship for Science and Technology.

Disclosure

The author reports no conflicts of interest.

References

1. Benos DJ, Saccomani G, Brenner BM, Sariban-Sohraby S. Purification and characterization of the amiloride-sensitive sodium channel from A6 cultured cells and bovine renal papilla. Proceedings of the National Academy of Sciences of the United States of America. 1986;83(22):8525–9.

2. Canessa CM, Horisberger JD, Schild L, Rossier BC. Expression cloning of the epithelial sodium channel. Kidney international. 1995;48(4): 950–5.

3. Canessa CM, Merillat AM, Rossier BC. Membrane topology of the epithelial sodium channel in intact cells. The American Journal of Physiology. 1994;267(6 pt 1):C1682–90.

4. Hummler E, Barker P, Gatzy J, Beermann F, Verdumo C, Schmidt A, et al. Early death due to defective neonatal lung liquid clearance in alpha-ENaC-defi cient mice. Nature Genetics. 1996;12(3):325–8.

5. Chraibi A, Verdumo C, Merillat AM, Rossier BC, Horisberger JD, Hummler E. Functional analyses of a N-terminal splice variant of the alpha subunit of the epithelial sodium channel. Cellular physiology and biochemistry: international journal of experimental cellular physiology, biochemistry, and pharmacology. 2001a;11(3):115–22.

6. Killick R, Richardson G. Isolation of chicken alpha ENaC splice variants from a cochlear cDNA library. Biochimica et biophysica acta. 1997;1350(1):33–7.

7. Li XJ, Xu RH, Guggino WB, Snyder SH. Alternatively spliced forms of the alpha subunit of the epithelial sodium channel: distinct sites for amiloride binding and channel pore. Molecular Pharmacology. 1995;47(6):1133–40.

8. Tucker JK, Tamba K, Lee YJ, Shen LL, Warnock DG, Oh Y. Cloning and functional studies of splice variants of the alpha-subunit of the amiloride-sensitive Na+ channel. The American Journal of Physiology. 1998;274(4 Pt 1):C1081–9.

9. Shehata MF. Characterization of Epithelial Sodium Channel Alpha Subunit Transcripts and Their Corresponding mRNA Expression Levels in Dahl S versus R. rats' Kidney Cortex On Normal and High Salt Diet. Proceedings from the 5th Northern Lights Conference, University of Waterloo, Waterloo, ON. 2007.

10. Husted RF, Takahashi T, Stokes JB. IMCD cells cultured from Dahl S rats absorb more Na+ than Dahl R rats. The American Journal of Physiology. 1996;271(5 Pt 2):F1029–36.

11. Shehata MF, Leenen FH, Tesson F. Sequence analysis of coding and 3' and 5' flanking regions of the epithelial sodium channel alpha, beta, and gamma genes in Dahl S versus R rats. BMC Genetics. 2007;8:35.

12. Chraibi A, Verdumo C, Merillat AM, Rossier BC, Horisberger JD, Hummler E. Functional analyses of a N.-terminal splice variant of the alpha subunit of the epithelial sodium channel. Cellular physiology and biochemistry: international journal of experimental cellular physiology, biochemistry, and pharmacology. 2001b;11(3):115–22.

13. Pandya MJ, Golderer G, Werner ER, Werner-Felmayer G. Interaction of human GTP cyclohydrolase I with its splice variants. The Biochemical Journal. 2006;400(1):75–80.

14. Valenzuela A, Talavera D, Orozco M, de la Cruz X. Alternative splicing mechanisms for the modulation of protein function: conservation between human and other species. Journal of Molecular Biology. 2004;335(2):495–502.

15. Zarei MM, Zhu N, Alioua A, Eghbali M, Stefani E, Toro L. A novel MaxiK splice variant exhibits dominant-negative properties for surface expression. The Journal of Biological Chemistry. 2001;276(19):16232–9.

A Novel Human NatA Nα-Terminal Acetyltransferase Complex: hNaa16p-hNaa10p (hNat2-hArd1)

Thomas Arnesen, Darina Gromyko, Diane Kagabo,
Matthew J. Betts, Kristian K. Starheim, Jan Erik Varhaug,
Dave Anderson and Johan R. Lillehaug

ABSTRACT

Background

Protein acetylation is among the most common protein modifications. The two major types are post-translational N^ε-lysine acetylation catalyzed by KATs (Lysine acetyltransferases, previously named HATs (histone acetyltransferases) and co-translational N^α-terminal acetylation catalyzed by NATs (N-terminal acetyltransferases). The major NAT complex in yeast, NatA, is composed of the catalytic subunit Naa10p (N alpha acetyltransferase 10 protein)

(Ard1p) and the auxiliary subunit Naa15p (Nat1p). The NatA complex potentially acetylates Ser-, Ala-, Thr-, Gly-, Val- and Cys- N-termini after Met-cleavage. In humans, the homologues hNaa15p (hNat1) and hNaa10p (hArd1) were demonstrated to form a stable ribosome associated NAT complex acetylating NatA type N-termini in vitro and in vivo.

Results

We here describe a novel human protein, hNaa16p (hNat2), with 70% sequence identity to hNaa15p (hNat1). The gene encoding hNaa16p originates from an early vertebrate duplication event from the common ancestor of hNAA15 and hNAA16. Immunoprecipitation coupled to mass spectrometry identified both endogenous hNaa15p and hNaa16p as distinct interaction partners of hNaa10p in HEK293 cells, thus demonstrating the presence of both hNaa15p-hNaa10p and hNaa16p-hNaa10p complexes. The hNaa16p-hNaa10p complex acetylates NatA type N-termini in vitro. hNaa16p is ribosome associated, supporting its potential role in cotranslational Nα-terminal acetylation. hNAA16 is expressed in a variety of human cell lines, but is generally less abundant as compared to hNAA15. Specific knockdown of hNAA16 induces cell death, suggesting an essential role for hNaa16p in human cells.

Conclusion

At least two distinct NatA protein Nα-terminal acetyltransferases coexist in human cells potentially creating a more complex and flexible system for Nα-terminal acetylation as compared to lower eukaryotes.

Background

About 80% of all mammalian proteins and 50% of yeast proteins are estimated to be cotranslationally acetylated at their N-termini [1-6]. This clearly makes N-terminal acetylation one of the most common protein modifications in eukaryotic cells. In yeast, three complexes, NatA, NatB and NatC, express different substrate specificities and are responsible for the majority of N-terminal acetylation [6]. At present, the nomenclature of this class of enzymes is not coherent and later this year a revised nomenclature of this enzyme class will be presented (Polevoda B, Arnesen T and Sherman F, unpublished). In brief, for the proteins mentioned in this study the following names will apply: Naa10p (Ard1), Naa11p (Ard2), Naa15p (Nat1), Naa16p (Nat2) and Naa50p (Nat5). The yeast NatA complex contains the structural subunit Naa15p mediating ribosome association and the catalytic subunit Naa10p [7,8]. Deletion of yNAA15 and yNAA10 results in a number of common defects including lack of Go entry, reduced cell growth, and inability to sporulate [9-11]. The subunit Naa50p is also physically associated with Naa10p and Naa15p, but the function of hNaa50p is unknown [7]. The human NatA, NatB and NatC complexes were recently characterized [12–15]. The human NatA complex contains the human homologues of the yeast NatA components hNaa10p, hNaa15p and hNaa50p [13,16]. The function and substrate specificity of hNatA in vivo and in vitro were found to resemble that of the yeast NatA complex [1]. The yeast NAA10 gene is duplicated in mammals. In humans, the NAA10 duplication has lead to the generation of a novel protein designated hNaa11p [17]. Similarly to hNaa10p, hNaa11p potentially interacts with hNaa15p implying that two distinct NatA complexes may exist in human cells: both hNaa15p-hNaa10p and hNaa15p-hNaa11p [17]. However, an endogenous hNaa15p-hNaa11p complex has not yet been detected, thus the functional importance of hNaa11p remains to be elucidated. hNaa10p and hNaa15p were previously demonstrated to be important for normal cellular viability. RNA interference-mediated knockdown of hNAA10 or hNAA15 induced apoptosis and cell cycle arrest in human cell lines [18-20], thus hNaa10p has been proposed to be a novel cancer drug target [21]. On the other hand, it has also been reported that hNaa10p is essential for the induction of apoptosis since knockdown of hNaa10p protected cells against doxorubicin induced apoptosis [22].

In order to identify novel interaction partners of hNaa10p, we performed immunoprecipitation of hNaa10p from HEK293 cells, followed by trypsin digestion of immunoprecipitated proteins and peptide analysis by mass spectrometry. We here demonstrate the existence of an endogenous hNaa16p protein, encoded by a human paralogue of the hNAA15 gene representing a new orthologue of the yeast NAA15 gene. hNAA16 mRNA is generally expressed in human cells. The hNaa16p protein associates with ribosomes, and interacts with hNaa10p to form a novel human NatA complex.

Results

Identification of hNaa16p and the hNaa16p-hNaa10p Complex

We used an immunoprecipitation-mass spectrometry approach to identify novel interaction partners of hNaa10p. Endogenous hNaa10p was collected from HEK293 cell extracts using a hNaa10p-specific antibody. The immunoprecipitates were analysed by LC/MS/MS after trypsin digestion. In addition to hNaa10p, hNaa15p and hNaa50p [13,16], we identified a novel protein, hNaa16p (hNat2/Entrez gene official symbol: NARG1L), in each of four parallel affinity extractions obtained using anti-hNaa10p. Both hNaa15p and hNaa16p were identified by several unique peptides. Neither hNaa10p nor hNaa15p nor hNaa16p were present in four negative controls using unspecific immunoglobulins. The identified hNaa15p and hNaa16p specific peptides are presented in Figure 1. The MS/MS spectra of two of these peptides, RAIELATTLDESLTNR (uniquely identifying hNaa15p) and DLESFNEDFLK (uniquely identifying hNaa16p) are shown in Figure 2. From the alignments (Figure 1) it is obvious that the two proteins hNaa15p and hNaa16p are highly similar. According to NCBI bl2seq [23], hNaa15p and hNaa16p share 70% identity and 85% similarity at the amino acid level. In Figure 3, hNaa15p and hNaa16p structural domains as identified by SMART [24] are presented. Both proteins contain several Tetratricopeptide (TPR) domains which are degenerate 34 amino acids repeats containing a helix-loop-helix presumed to be involved in protein-protein interactions. hNaa15p is predicted to contain 4 TPRs while hNaa16p is predicted to have 5 TPRs. This difference results from program threshold values chosen, and since both proteins are highly similar in this region, it is likely that also hNaa15p may have a fifth TPR domain in the same region. Both proteins contain coiled-coil domains (Figure 3). Also the exon-intron organisation for the two genes is highly similar (Figure 3).

```
hNaa15p    1   MPAVSLPPKENALFKRILRCYEHKQYRNGLKFCKQILSNPKFAEHGETLAMKGLTLNCLG 60
               MP V LPPKE+ LFKRIL+CYE KQY+NGLKFCK ILSNPKFAEHGETLAMKGLTLNCLG
hNaa16p    1   MPNVLLPPKESNLFKRILKCYEQKQYKNGLKFCKMILSNPKFAEHGETLAMKGLTLNCLG 60

hNaa15p   61   KKEEAYELVRRGLRNDLK SHVCWHVYGLLQRSDKKYDEAIKCYRNALKWDK DNLQILR DL 120
               KKEEAYE VR+GLRND+K SHVGWHIVVGLLQRSDRKYDEAIKCYRNALK DK DNLQILR DL
hNaa16p   61   KKEEAYEFVRKGLRNDVK SHVCWHVYGLLQRSDKKYDEAIKCYRNALKLDK DNLQILR DL 120

hNaa15p  121   SLLQIQMR DLEGYRETRYQLLQLRPAQR ASWIGYAIAYHLLEDYEMAAKILEEFRKTQQT 180
               SLLQIQMR DLEGYRETRYQLLQLRP QR ASWIGYAIAYHLL+DY+MA K+LEEFR+TQQ
hNaa16p  121   SLLQIQMR DLEGYRETRYQLLQLRPTQR ASWIGYAIAYHLLKDYDMALKLLEEFRQTQQV 180

hNaa15p  181   SPDKVDYEYSELLLYQNQVLREAGLYREALEHLCTYEKQICDK LAVEETKGEILLQLCR L 240
               P+K+DYEYSEL+LYQNQV+REA L +E+LEH+ YEKQICDK L VEE KGE+LL+R L
hNaa16p  181   PPNKIDYEYSELILYQNQVMREADLLQESLEHIEMYEKQICDK LLVEEIKGEILLKLGR L 240

hNaa15p  241   EDAADVYRGLQERNPENWAYYKGLEK ALKPANMLERLKIYEEAWTKYPRGLVPRRLPLNF 300
               ++A++V++ L +RN ENW YY+GLEK AL+ + + ERL+IYEE  ++P+ + PRRLPL
hNaa16p  241   KEASEVFKNLIDRNAENWCYYEGLEK ALQISTLEERLQIYEEISKQHPKAITPRRLPLTL 300

hNaa15p  301   LSGEKFKECLDKFLRMNFSKGCPPVFNTLRSLYKDKEKVAIIEELVVGYETSLKSCRLFN 360
               + GE+F+E +DKFLR+NFSKGCPP+F TL+SLY + EKV+II+ELV  YE SLK+C F+
hNaa16p  301   VPGERFRELMDKFLRVNFSKGCPPLFTTLKSLYYNTEKVSIIQELVTNYEASLKTCDFFS 360

hNaa15p  361   PNDDGKEEPPTTLLWVQYYLAQHYDKIGQPSIALEYINTAIESTPTLIELFLVKAKIYKH 420
               P ++G++EPPTTLLWVQY+LAQH+DK+GQ S+AL+YIN AI STPTLIELF +KAKIYKH
hNaa16p  361   PYENGEKEPPTTLLWVQYFLAQHFDKLGQYSLALDYINAAIASTPTLIELFYMKAKIYKH 420

hNaa15p  421   AGNIKEAARWMDEAQALDTADRFINSKCAKYMLKANLIKEAEEMCSKFTREGTSAVENLN 480
               GN+KEAA+WMDEAQ+LDTADRFINSKCAKYML+AN+IKEAEEMCSKFTREGTSA+ENLN
hNaa16p  421   IGNLKEAAKWMDEAQSLDTADRFINSKCAKYMLRANMIKEAEEMCSKFTREGTSAMENLN 480

hNaa15p  481   EMQCMWFQTECAQAYKAMNKFGEALKKCHEIERHFIEITDDQFDFHTYCMRKITLRSYVD 540
               EMQCMWFQTEC  AY+ + ++G+ALKKCHE+ERHF  ITDDQFDFHTYCMRK+TLR+YVD
hNaa16p  481   EMQCMWFQTECISAYQRLGRYGDALKKCHEVERHFFEITDDQFDFHTYCMRKMTLRAYVD 540

hNaa15p  541   LLKLEDVLRQHPFYFKAARIAIEIYLK LHDNPLTDENKEHEADTANMSDKELKKLRNKQR 600
               LL+LED+LR+H FYFKAAR AIEIYLK L+DNPLT+E+K+ E ++ N+S KELKK+ +KQR
hNaa16p  541   LLRLEDILRRHAFYFKAARSAIEIYLK LYDNPLTNESKQQEINSENLSAKELKKMLSKQR 600

hNaa15p  601   RAQKKAQIEEEKKNAEKEKQQRNQKKKDDDDEEIGGPKEELIPEK LAKVETPLEEAIKF 660
               RAQKKA++EEE+K+AE+E+QQ+NQKKK+D+++EE  G KEELIPEK L +VE PLEEA+KF
hNaa16p  601   RAQKKAKLEEERKHAERERQQKNQKKKRDEEEEEASGLKEELIPEK LERVENPLEEAVKF 660

hNaa15p  661   LTPLKNLVKNKIETHLFAFEIYFRKEKFLLMLQSVKRAFAIDSSHPWLHECMIR LFNTAV 720
               L PLKNLV + I+THL AFEIYFRK KFLLMLQSVKRAFAI+S++PWLHEC+IR  F+ +V
hNaa16p  661   LIPLKNLVADNIDTHLLAFEIYFRKGKFLLMLQSVKRAFAINSNNPWLHECLIR  −FSKSV 719

hNaa15p  721   CESKDLSDTVRTVLRQEMNRLFQEFNPKNFNETFLK RNSDSLPHRLSAAKMVYYLDPSSQ 780
               +L D V  VL QEM ++F   + ++FNE FLK RN+ SL H LS AKM+Y+LD S Q
hNaa16p  720   SNHSNLPDIVSKVLSQEMQKIFVKKDLESFNEDFLK RNATSLQHLLSGAKMMYFLDKSRQ 779

hNaa15p  781   K RAIELATTLDESLTNR NLQTCMEVLEALYDGSLDGHEAAEIYRANCHKLFPYALAFMP 840
               + +AI +AT LDE++ ++ +++T +V EAL DGS G+C   E YR  CH L P+  AF+P
hNaa16p  780   E KAIAIATRLDETIKDK DVKTLIKVSEALLDGSFGNCSSQYEEYRMACHNLLPFTSAFLP 839

hNaa15p  841   PGYEEDMK ITVNGDSSAEAEELANEI 866
               E D      V  +A  + LANEI
hNaa16p  840   AVNEVD−N PNVALNHTANYDVLANEI 864
```

Figure 1. Alignment of hNaa15p and hNaa16p proteins with identified peptides from anti-hNaa10p affinity extracts indicated. Sequences for hNaa15p and hNaa16p were aligned using Muscle [36], with sequences common to both proteins shown between the hNaa15p and hNaa16p sequences. Tryptic peptides uniquely identifying hNaa15p are in bold and red, sequences uniquely identifying hNaa16p are in bold and green, and peptides common to both proteins are bold and black. Breaks indicate two different peptides within a region. A number of peptides unique to each protein confirmed that hNaa16p has a different sequence from hNaa15p.

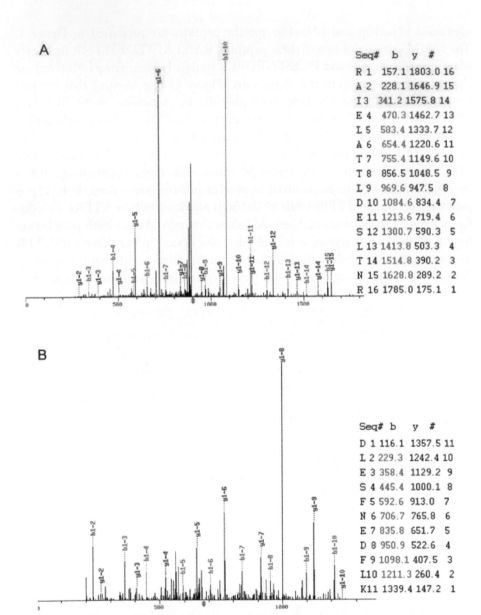

Figure 2. MS/MS spectra of peptides uniquely identifying hNaa15p and hNaa16p in complex with hNaa10p. A. MS/MS spectrum of the peptide RAIELATTLDESLTNR, which uniquely identifies hNaa15p in the hNaa10p affinity extraction. The MS/MS spectrum of the +2 peptide ion was matched to this sequence with an Xcorr of 4.25 [30]; the peptide had a probability of correct sequencing of 0.99 [32]. B. MS/MS spectrum of the peptide DLESFNEDFLK, which uniquely identifies hNaa16p in the hNaa10p affinity extraction. The MS/MS spectrum of the +2 peptide ion was matched to this sequence with an Xcorr of 3.4; the peptide had a probability of correct sequencing of 0.95. All fragment ions have a +1 charge. Individual matched y ions (blue) and b ions (red) are indicated in bold; the position of the peptide precursor ion is indicated by the dagger; the x-axis units are m/z and the y-axis represents relative peak intensity normalized to the most intense fragment ion.

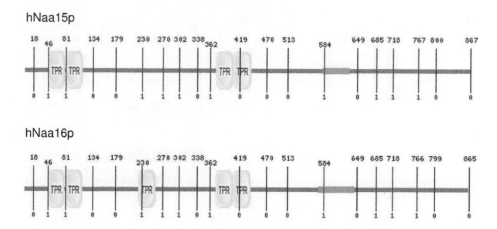

Figure 3. Overview of structural domains and exon structure of hNaa15p and hNaa16p. SMART analysis [24] demonstrates the presence of TPR domains (yellow) and coiled coil regions (green). Intron positions are indicated with vertical lines showing the intron phase (below lines) and exact amino acid position (above lines).

The interaction between hNaa16p and hNaa10p appears to be independent of the hNaa15p-hNaa10p complex since hNaa16p-specific peptides were not present in anti-hNaa15p immunoprecipitates analysed in parallel. Since both hNaa15p and hNaa16p interact with hNaa10p, it is of interest to determine the approximate ratio of hNaa15p versus hNaa16p in complex with hNaa10p. To make a rough and qualitative estimate of the amounts of hNaa15p and hNaa16p complexed with hNaa10p, the protein abundance index [25] was calculated for each protein as the number of LC/MS/MS-observed tryptic peptides (unique to each protein) divided by the total number of tryptic peptides that could be identified by SEQUEST under the exact search conditions used. For hNaa15p, this number was 0.0858, while for hNaa16p it was 0.0138. hNaa15p-hNaa10p complexes thus appear to be roughly 6-fold more abundant than hNaa16p-hNaa10p complexes in HEK293 cells. As an additional verification of the hNaa16p-hNaa10p interaction, we expressed tagged hNaa16p-FLAG and hNaa10p-V5 in HeLa cells and demonstrated that hNaa16p-FLAG was co-immunoprecipitated with hNaa10p-V5 (Figure 4A). In conclusion, these observations clearly support the presence of an endogenous hNaa16p-hNaa10p complex in human cells. Furthermore, we also wanted to test whether hNaa16p could interact with hNaa11p, since the latter protein is very similar to hNaa10p. The co-immunoprecipitation experiments using hNaa16p-FLAG and hNaa11p-V5, demonstrated that this indeed is the case (Figure 4B).

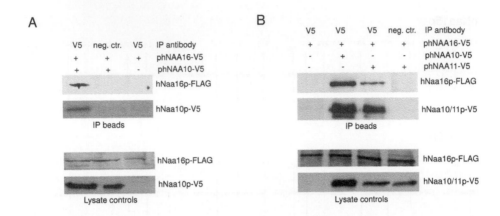

Figure 4. hNaa10p-V5 and hNaa11p-V5 co-immunoprecipitate hNaa16p-FLAG. (A) HeLa cells were transfected with phNAA10-V5 and phNAA16-FLAG as indicated and the cell lysates were immunoprecipitated (IP) with anti-V5 or unspecific control antibodies. The immunoprecipitates and lysate controls were analyzed by SDS-PAGE and Western blotting using anti-FLAG and anti-V5 antibodies. The amount of the lysate loaded on the gel represents approximately 20% of the material used for IP. Results are representative of three independent experiments. (B) As (A) using phNAA11-V5 and phNAA16-FLAG plasmids.

Expression of hNAA15 and hNAA16 in Human Cell Lines

To investigate the expression of these two human homologues of the yeast NAA15 gene, we analysed the expression of hNAA15 and hNAA16 mRNA by quantitative RT-PCR in 10 human cell lines of various origin (Figure 5A) and 10 human cell lines originating from the thyroid gland (Figure 5B). Both genes are highly expressed in HepG2 (hepatocellular carcinoma), HEK293 (kidney adenocarcinoma), and SK-MEL2 (malignant melanoma) cell lines. Our results demonstrate that hNAA16 is expressed in all cell lines analyzed, and hNAA15 is the most abundant species of the two. The expression level of hNAA15 mRNA is 2–3 times than that of hNAA16 mRNA in cell lines like TAD-2 and HepG2, while it is 7–11 times more abundant in GaMG, SK-N-MC, HeLa, and B-CPAP cells. Interestingly, in HEK293 cells, hNAA15 mRNA is 5 times more abundant as compared to hNAA16 mRNA. Although there is not necessarily a direct relationship between the level of a specific mRNA and the level of the corresponding protein, this agrees with our rough estimates that hNaa15p-hNaa10p complexes appear to be 6-fold more abundant than hNaa16p-hNaa10p complexes in HEK293 cells, thus suggesting that the level of the various NatA complexes present is proportional to the presence of their single subunits. Tissue expression profiles for hNAA15 and hNAA16 mRNA also indicate that hNAA15 mRNA is generally more abundant as compared to hNAA16 (NCBI-UNIGENE-EST PROFILE VIEW-ER). However, in certain tissues like adrenal gland, mammary gland, heart, testis and thymus, hNAA16 appears to be the dominant species (http://www.ncbi.nlm.nih.gov/UniGene/ESTProfileViewer.cgi?uglist=Hs.715706 and http://www.ncbi.nlm.nih.gov/UniGene/ESTProfileViewer.cgi?uglist=Hs.512914). Previously, it was demonstrated that hNAA15 was overexpressed in papillary thyroid carcinomas as compared to non-neoplastic thyroid tissue [26,27]. The current data support these observations since hNAA15 is slightly or significantly overexpressed in all types of thyroid cancer cell lines tested as compared to the primary thyroid cells (TAD-2/Nthy ori 3.1): follicular thyroid carcinoma (CGTH-W-1), papillary thyroid carcinoma (NPA/B-CPAP/ONCO-DG-1/BHT-101) and anaplastic thyroid carcinoma (8305C/CAL-62/ARO) (Figure 5B). On the other hand, we are not able to make similar conclusions with respect to hNAA16 which appears to be expressed equally in tumour- and non-tumour thyroid cell lines.

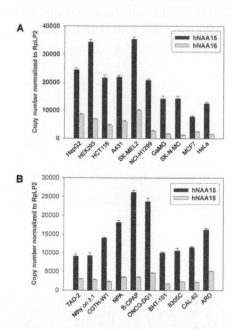

Figure 5. Expression of hNAA15 and hNAA16 in human cell lines. (A) The expression of hNAA15 and hNAA16 mRNA in 10 different human cell lines was analysed by real-time quantitative PCR (RT-qPCR) using gene specific primers. Values were normalized to the expression of RpLP2. An independent normalization was performed using a second reference gene, RPol2, producing almost identical results (data not shown). The mean ± SD of triplicate experiments was calculated. HepG2, hepatocellular carcinoma; HEK293, kidney adenocarcinoma; HCT116, colon carcinoma; A431, epidermoid carcinoma; SK-MEL-2, malignant melanoma; NCI-H1299, non-small cell lung carcinoma; GaMG, glioblastoma; SK-N-MC, neuroepithelioma; MCF7, breast adenocarcinoma; HeLa, cervix adenocarcinoma. (B) As (A) using 10 different human cell lines originating from the thyroid gland. TAD-2/Nthy ori 3.1, primary human thyroid cells; CGHT-W-1, follicular thyroid carcinoma; NPA/B-CPAP/ONCO-DG-1/BHT-101, papillary thyroid carcinoma; 8305C/CAL-62/ARO, anaplastic thyroid carcinoma.

Evolution of the hNAA16 Gene

The hNAA15 gene (Entrez gene official symbol: NARG1) is located to chromosome 4 (4q31.1) while the hNAA16 gene (Entrez gene official symbol: NARG1L) is located to chromosome 13 (13q14.11). hNaa15p and hNaa16p belong to a protein family with members in Saccharomyces cerevisiae, Caenorhabditis elegans, Ciona intestinalis, Drosophila melanogaster, fish, frog and several higher mammals (Family ENSF00000002142 from Ensembl v38, April 2006 [see methods]). The phylogenetic tree of this family (Figure 6) suggests that the duplication that resulted in hNAA15 and hNAA16 occurred at some point after the speciation that resulted in the higher chordates (i.e. after the divergence of Ciona), but before the speciation of mammals into eutheria and metatheria. The exact timing is confused by Ciona branching before Drosophila, and by the fish (zebrafish, fugu, tetraodon) appearing in only the NAA15 clade (possibly suggesting loss of NAA16 in fish). The duplication therefore occurred before the duplication that resulted in hNAA10 and hNAA11, which we previously narrowed down to some time shortly after the divergence of eutheria and metatheria [17].

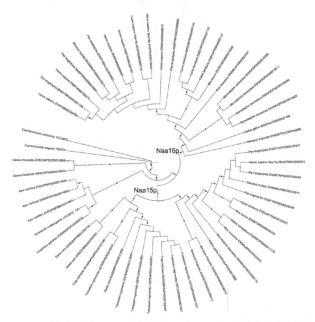

Figure 6. Phylogenetic tree of NAA15 genes. A phylogenetic tree showing that the duplication leading to hNAA15 and hNAA16 occurred at some point after the divergence of Ciona from the other chordates, and before the speciation that resulted in the amphibia. This is before the duplication resulting in the NAA10s [17]. All identifiers are from Ensembl. Branches highlighted with a black dot are those where the probability of the split represented by the branch is > 0.95. Probable NAA15 and NAA16 clades are indicated. In-paralogues in this tree represent alternative transcripts, which are possibly artefacts of the genome annotation. See Methods for details.

hNaa16p Associates with Ribosomes and hNaa16p-hNaa10p Acetylates NatA-Type N-Termini In Vitro

The characterized hNaa15p-hNaa10p hNatA complex interacts with ribosomes [13] and N-terminally acetylates polypeptides with Ser-, Ala-, Thr-, Val- and Gly- N-termini [1]. Analyses of isolated polysomes demonstrated that hNaa16p-FLAG, as well as hNaa15p and hNaa10p, were both present in the polysomal and the soluble fractions (Figure 7) supporting a model where hNaa16p dynamically associates with ribosomes, and is involved in cotranslational Nα-terminal acetylation. To assess whether the hNaa16p-hNaa10p complex expresses acetyltransferase activity, we performed in vitro N-terminal acetyltransferase assays using immunoprecipitated hNaa16p-hNaa10p and synthetic oligopeptides as substrates. A synthetic oligopeptide representing an acetylated protein N-terminus in HeLa cells, SESS-, was acetylated in vitro by hNaa16p-hNaa10p, while an oligopeptide representing an N-terminus not acetylated in HeLa cells, SPTP-, carrying an inhibitory Pro in the second position [1], is not efficiently acetylated in vitro (Figure 8). These results suggest that hNaa16p-hNaa10p and hNaa15p-hNaa10p have overlapping substrate specificities, since also the hNaa15p-hNaa10p complex showed a similar in vitro reactivity towards these peptides [1]. It should be noted, however, that in order to properly define the substrate specificity of the hNaa16p-hNaa10p complex, a more comprehensive approach is required.

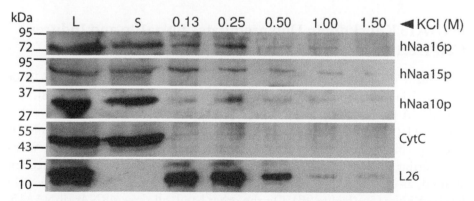

Figure 7. hNaa16p co-sediments with polysomal fractions in a salt sensitive manner. Polysomal pellets from HeLa cells expressing hNaa16p-FLAG were resuspended in buffer containing increasing concentrations of KCl. Cell lysate (L), supernatant post first ultracentrifugation (S) and polysomal pellets after KCl treatment were analyzed on SDS-PAGE/Western blotting. The membrane was incubated with anti-FLAG, anti-hNaa15p, anti-hNaa10p, anti-L26 (ribosomal protein) and anti-CytC antibodies. Molecular-mass markers (in kDa) are indicated on the left hand side. Results shown are representative of three independent experiments.

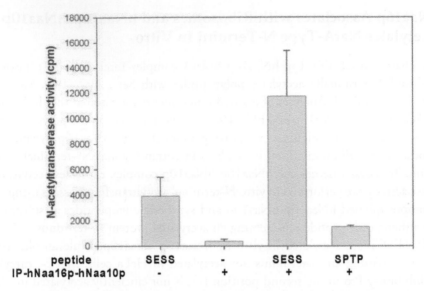

Figure 8. N-terminal acetyltransferase activity of the hNaa16p-hNaa10p complex. HeLa cells were transfected with phNAA16-FLAG and phNAA10-V5, and immunoprecipitated with anti-FLAG (+) or a negative control antibody (-). The immunoprecipitated hNaa16p-hNaa10p complex was subjected to an in vitro N-acetylation assay by the addition of acetylation buffer, a synthetic oligopeptide, [14-C]-Acetyl coenzyme A. The radioactivity incorporated into the peptide was determined by scintillation counting. The peptides used are [H]SESSSKSRWGRPVGRRRRPVRVYP [OH] (SESS) and [H]SPTPPLFRWGRPVGRRRRPVRVYP [OH] (SPTP). An equal level of immunoprecipitated material (input enzyme) was verified by Western-blotting analysis. Results are presented as mean ± SD from three independent experiments.

hNAA16 Knockdown Downregulates hNaa10p and Induces Cell Death

Previously, we demonstrated that siRNA-mediated knockdown of hNAA10 or hNAA15 induced apoptosis and cell cycle arrest in human cell lines [18]. Using siRNA pools specific for hNAA15 and hNAA16, we knocked down either gene in HeLa cells. Semiquantitative RT-PCR analysis revealed that both siRNA pools specifically reduced the expression of the targeted gene (Figure 9A). Western blotting analysis of cell lysates harvested 72 hours post siRNA-transfection showed, as expected, that only sihNAA15, not sihNAA16, reduced the protein levels of hNaa15p (Figure 9B). On the other hand, all three siRNAs sihNAA15, sihNAA16 and sihNAA10 significantly reduced protein levels of hNaa10p. The dependency of hNaa10p for hNaa15p has previously been described [26,28], but the present results also indicate that a fraction of endogenous hNaa10p depend on hNaa16p for stability, albeit to a lesser extent as compared hNaa15p. This is in agreement with the relative presence of hNaa15p-hNaa10p versus hNaa16p-hNaa10p complexes described above.

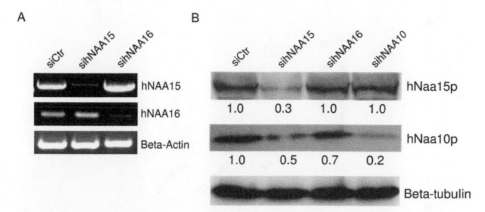

Figure 9. Knockdown of hNAA16 downregulates hNaa10p. (A) HeLa cells cultured in 6 cm dishes were transfected with 50 nM unspecific (Ctr), hNAA15 or hNAA16 siRNAs. ZVAD-fmk (5 μM) was added 24 hours post transfection to prevent any induction of apoptosis. After 48 hours total RNA was isolated and processed by RT-PCR with specific primers to hNAA15, hNAA16 and β-actin. (B) HeLa cells cultured in 6 cm dishes were transfected with 50 nM unspecific (Ctr), hNAA15, hNAA16 or hNAA10 siRNAs. ZVAD-fmk (5 μM) was added 24 and 48 hours post transfection to prevent any induction of apoptosis. After 72 hours cell lysates were analyzed by SDS-PAGE/Western blotting. The membrane was incubated with anti-hNaa10p, anti-hNaa15p and anti-β-tubulin. Results shown are representative of three independent experiments. Protein levels were quantified using FUJIFILM IR LAS 1000 and Image Gauge 3.45. Protein levels in siCtr (-) samples were set to 1.0 and protein levels in sihNAA15, sihNAA16 or sihNAA10 treated cells were estimated relative to this and normalized to β-tubulin levels.

Knockdown of hNAA10 and hNAA15 induces apoptosis in HeLa cells [18], and in the case of these cells, we demonstrated that hNAA16 knockdown reduced cell viability at comparable levels to hNAA15 knockdown as determined by a WST-1 assay (Figure 10A). In order to investigate this phenotype, we analysed knockdown cells by live microscopy and observed the appearance of dead cells: cells detaching from monolayer had characteristic features of apoptosis (condensation of nucleus, shrinking of cytoplasm, formation of apoptotic bodies). Furthermore, by Hoechst staining we observed a significant increase of pycnotic nuclei in hNAA15 and hNAA16 knockdown cells as compared to control cells (Figure 10B). A similar observation was made using a TUNEL assay to detect DNA fragmentation further supporting that knockdown of hNAA16 induces cell death, most likely apoptosis (Figure 11).

Discussion

Phylogenetic analysis provides strong support for a vertebrate gene duplication resulting in the two copies of the NAA15 gene, possibly followed by loss of the second copy in fish. The presence of hNaa16p specific peptides in four of four parallel anti-hNaa10p immunoprecipitates is a strong indication of the presence

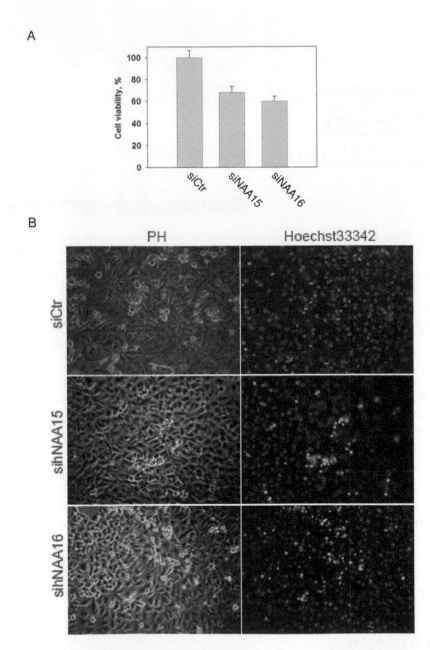

Figure 10. Knockdown of hNAA16 reduces cell viability. (A) HeLa cells were transfected with 50 nM unspecific, hNAA15 or hNAA16 siRNAs. Cell viability was determined by WST-1 assay 72 hours post transfection. The viability of the control cells transfected with Ctr siRNA was set to 100%. The mean ± SD of three parallel experiments was calculated. (B) HeLa cells were transfected with 50 nM unspecific, hNAA15 or hNAA16 siRNAs for 72 hours and analyzed by phase contrast microscopy (PH) to observe cell morphology and Hoechst33342 staining to visualize nuclei.

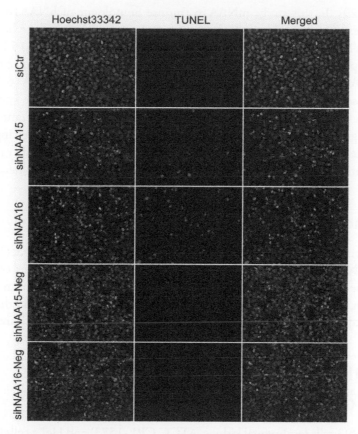

Figure 11. Knockdown of hNAA16 induces cell death. HeLa cells were transfected with 50 nM unspecific, hNAA15 or hNAA16 siRNAs. Cells were analyzed by TUNEL-assay 72 hours post transfection. In the left column, Hoechst33342 staining visualizes nuclei in blue; in the middle column, TMR-red TUNEL staining visualizes DNA breaks in red; in the right column, Hoechst and TUNEL signals are merged. Neg: negative controls omitting active TUNEL-reaction enzyme

of endogenous hNaa16p protein and of the presence of hNaa16p-hNaa10p complexes. Both hNaa15p and hNaa16p are orthologues of the yeast Naa15p. The simultaneous presence of hNaa15p and hNaa16p in anti-hNaa10p immunoprecipitates, and the lack of hNaa16p in anti-hNaa15p immunoprecipitates, indicate that hNaa15p-hNaa10p and hNaa16p-hNaa10p make distinct NatA complexes in the cells. Recently, we also demonstrated the presence of a second human orthologue of the yeast NAA10 gene [17]. Thus, two human orthologues of the yeast Naa10p protein exist, hNaa10p and hNaa11p. However, to date there is no evidence of the presence of endogenous complexes consisting of hNaa11p and hNaa15p or hNaa16p. In mouse, two homologues of yeast NAA15, mNAA15 and mNAA16 (named mNAT1 and mNAT2), were detected at the mRNA level [29]. In this study, mNAA16 (mouse homologue of hNAA16) was found to be expressed at a lower level as compared to mNAA15 (mouse homologue of hNAA15), in agreement with our results suggesting that hNAA15 is the dominant species as compared to hNAA16. Despite the lower abundance of hNAA16 mRNA as compared to hNAA15 mRNA, we find that hNAA16 is generally expressed in a number of human cell lines (Figure 5) and that knockdown of hNAA16 induces apoptosis (Figure 11), thus suggesting an important role for this novel gene. Whether the hNAA16 knockdown phenotype is resulting from lack of N-terminally acetylated hNaa16p-hNaa10p substrates or whether it is coupled to another unknown function of hNaa16p remains to be investigated.

Conclusion

In summary, we have identified hNaa16p, a novel hNaa10p interactor. hNaa16p and hNaa10p most likely represent a novel human NatA complex complementing the role of the hNaa15p-hNaa10p complex in protein Nα-terminal acetylation in human cells. The fact that hNAA16 is widely expressed and that hNAA16 knockdown induces apoptosis points to hNaa16p as an essential human protein.

Methods

Cell Culture and Transfection

HEK293 cells (embryonal kidney, ATCC CRL-1573) and HeLa cells (human cervix carcinoma, DSMZ no. ACC 57) were cultured at 37°C, 5% CO2 in DMEM supplemented with 10% FBS and 3% l-glutamine. Plasmid transfection was performed using Fugene6 (Roche) according to the manufacturers instructions. Plasmid expressing hNaa16p (NARG1L)-FLAG was purchased from Origene (RC214224; NM_024561). Plasmids expressing hNaa10p-V5 and hNaa11-V5 have been described [13,17]. siRNA transfection was performed using Dharmafect (Dharmacon) according to the instruction manual. Gene specific smart pool siRNAs were purchased from Dharmacon and used at a final concentration of 50 nM to silence the hNAA15 (hNAT1/NATH/NARG1) and hNAA16 (hNAT2/NARG1L) genes: sihNAA15, ON-TARGET plus SMARTpool, Cat. L-012847; sihNAA16, ON-TARGET plus SMARTpool, Cat. L-013336; negative control (siCtr), ON-TARGETplus Non-Targeting Pool, Cat. no D-001810.

For TUNEL assay, cytospins were prepared and cells were incubated in TUNEL reaction mixture (In Situ Cell Death Detection Kit, TMR red, Roche) for 1 hour at 37°C. Negative controls for hNAA15 and hNAA16 knockdown cells were incubated in TUNEL Label reaction without enzyme.

Immunoprecipitation and Mass Spectrometry

Immunoprecipitation and Mass Spectrometry were essentially performed as previously described [13], but omitting the crosslinking of antibodies to Protein AG Agarose beads. Rabbit polyclonal peptide specific antibodies against hNaa15p (anti-NATH) and hNaa10p (anti-hARD1) were produced by Biogenes GmBH. The immunogens correspond to amino acids 853–866 of hNaa15p and amino acids 204–217 of hNaa10p. Anti-hNaa10p does not crossreact with hNaa11p since these two proteins are significantly different in the antibody binding region [17]. A Thermofinnigan LCQ Deca XP Plus and an Agilent 1100 Nanoflow hplc were used for microcapillary LC/MS/MS analysis of tryptic peptides. Data analysis utilized SEQUEST [30], Medusa [31] and support vector machine learning [32]. Approximately 5×10^7 HEK293 cells were used per sample. Four samples were immunoprecipitated in parallel using anti-hNaa15p, four samples using anti-hNaa10p and four samples were immunoprecipitated using rabbit Ig (DAKO) as negative controls. Peptides present in the negative controls were subtracted from the peptides present in the anti-hNaa15p and anti-hNaa10p samples using Medusa [31].

Immunoprecipitation and Western Blotting

Approximately 2×10^6 cells were harvested for each sample. Cells were lysed in 300 μl IPH lysis buffer (50 mM Tris pH 8, 50 mM NaCl, 0.5% NP-40, 5 mM EDTA, 1 mM Na_3VO_4, 1 mM Pefabloc (Roche)) and incubated for 5 minutes on ice. Cell membranes and organelles were removed by centrifugation at 15700 × g for 30 seconds, and the cell lysate was transferred to a new tube. To remove proteins that bind unspecifically to the agarose beads, Protein A/G Agarose (Santa Cruz) was added, and the lysate was incubated on a roller for 30 minutes at 4°C. The beads were removed by centrifugation at 1500 × g for 4 minutes. The cell lysate was then incubated with 2 μg specific antibody on a roller for 1 – 4 hours at 4°C, before adding 50 μl Prot A/G Agarose beads. The lysate was then incubated on a roller at 4°C for 4 – 10 hours. The beads were collected by centrifuging as above, and washed with 1 × PBS. The supernatant was removed, and the samples were prepared for analysis by SDS-PAGE and Western Blotting as described [21]. Polyclonal rabbit antibodies against hNaa10p and hNaa15p described previously [13], were used at 1:500 dilution, anti-V5 (Invitrogen) at 1:2000, anti-FLAG (Sigma) at 1:2000. Horseradish peroxidase-linked anti-mouse and anti-rabbit were from Amersham Biosciences (Little Chalfont, Bucks., U.K.).

In Vitro Nᵃ-Acetyltransferase Assay

HEK293 cells were transfected by plasmids as described above and indicated in Figure 8, harvested and lysed in 300 µl IPH lysis buffer. Typically, 5×10^6 cells were used. 40 µl Protein A/G Agarose (Santa Cruz) was added to the lysates and incubated for one hour at 4°C. After centrifugation at $1500 \times g$ for 2 min, the supernatants were collected and incubated for another 2 hours at 4°C with anti-FLAG or unspecific antibody (2 µg). The samples were centrifuged as above and 50 µl Protein A/G Agarose was added to the supernatants. After incubation for 16 hours, centrifugation and three times of washing in $2 \times$ PBS and once in acetylation buffer (50 mM Tris-HCl, pH 8.5, 1 mM DTT, 800 µM EDTA, 10 mM Na-butyrate, 10% Glycerol), the samples were subjected to an in vitro acetylation assay. 10 µl peptide (0.5 mM, custom made peptides from Biogenes), 4 µl [¹⁴C] Acetyl-CoA (50 µCi, 2.07 GBq/mmol, GE Healthcare) and 250 µl acetylation buffer was added to pellets of Protein A/G-Agarose bound hNaa16p-FLAG-hNaa10p-V5 complexes. The mixture was incubated for 2 hours at 37°C with rotation. After centrifugation the supernatant was added to 250 µl SP Sepharose (50% slurry in 0.5 M acetic acid, Sigma) and incubated on a rotor for 5 min. The mixture was centrifuged and the pellet was washed three times with 0.5 M acetic acid and finally with methanol. Radioactivity in the peptide-containing pellet was determined by scintillation counting. All custom made peptides contains 7 unique amino acids at the N-terminus, since these are the major determinants for N-terminal acetylation. The next 17 amino acids are identical to the ACTH peptide (corticotrophin amino acid 1–24) sequence to maintain a positive charge facilitating peptide solubility and effective isolation by cation exchange Sepharose beads. The ACTH derived lysines were replaced by arginines to minimize any potential interference by Nε-acetylation. Peptide sequence information: High mobility group protein A1 (P17096): [H]SESSSKSRWGRPVGRRRRPVRVYP [OH], THO complex subunit 1 (Q96FV9): [H]SPTPPLFRWGRPVGRRRRPVRVYP [OH].

Isolation of Polysomes

Total ribosome isolation was performed as a modification of previously described methods [33,34]. Approximately 2×10^7 HEK293 cells were used per experiment. Prior to harvesting, cells were treated with 10 µg/ml cycloheximide (CHX) for 5 minutes at 37°C. Cells were harvested, and lysed with KCl ribosome lysis buffer (1.1% (w/v) KCl, 0.15% (w/v) triethanolamine, 0.1% (w/v) magnesium acetate, 8.6% (w/v) sucrose, 0.05% (w/v) Na-Deoxycholate, 0.5% (v/v) Triton-X100, 0.25% (v/v) Pefabloc), and incubated on ice for 15 minutes. After removing nucleus and membranes by centrifugation at $400 \times g$ for 10 min, 700 µl cell lysate was ultracentrifuged at 436,000 × g for 25 minutes on a 0.4 ml pillow of 25% sucrose in KCl ribosome lysis buffer using a MLA-130 rotor (Beckman, Geneva, Switzerland). Pellets were resuspended in ribosome lysis buffer with the indicated KCl concentrations, followed by ultracentrifugation as described above. Pellets were resuspended in KCl ribosome lysis buffer, and prepared for analysis by SDS-PAGE and Western blot.

RNA Purification and cDNA Synthesis

Total RNA was extracted with TRIzol reagent (Invitrogen, San Diego, CA) according to manufacturer's instructions. RNA was subsequently dissolved in DEPC-treated double-distilled water. Single-strand cDNA was synthesized from 1 μg total RNA using Transcriptor Reverse Transcriptase (Roche, Indianapolis, IN) and oligo(dT)15 primer according to the manufacturer's instructions.

Real-Time Quantitative PCR

Relative gene expression levels of hNAA15 and hNAA16 in thyroid cell lines and cell lines from different tissues were determined by real-time quantitative PCR (RT-qPCR).

To amplify hNAA15 and hNAA16 cDNA the following primers were used: hNAA15 primer set 1 (forward TTGGCACGTTTATGGCCTTCT and reverse CGTTTCCCTGTAACCCTCAAGA), hNAA15 primer set 2 (forward TGTATGGAGGTATTGGAAGCC and reverse CTCTTCATATCCAGGAGGCAT); hNAA16 primer set 1 (forward TCTTCCAGACATTGTGAGCAAAG and reverse AGGTAGCGTTACGTTTCAGAAAA) and hNAA16 primer set 2 (forward CAAGATGATTCTGTCGAACCCA and reverse AACGCTGCAAGAGTCCATATAC). For data normalization two reference genes were used independently: large ribosomal protein P2, RpLP2 (GeneID: 6181) and RNA polymerase II, RPol2 (GeneID: 5430). Primers for amplification of reference genes were as following: RpLP2 (forward GACCGGCTCAACAAGGTTAT and reverse CCCCACCAGCAGGTACAC); RPol2 (forward GCACCACGTCCAATGACAT and reverse GTGCGGCTGCTTCCATAA). Templates (equal amount of cDNA) and primers were mixed with components from the LightCycler 480 SYBR Green I Master mix kit (Roche Applied Science). Reactions in triplicate were carried out in the LightCycler 480 real-time PCR machine (Roche Applied Science) under the following conditions: initial denaturation at 95°C for 5 min, and then 40 cycles of denaturation at 95°C for 10 s, annealing at 57°C for 10 s, and extension at 72°C for 10 s. Melting curves were obtained to examine the purity of amplified products. Absolute quantitation data and CP values were obtained by analysis with LightCycler 480 Software 1.5 by 2nd derivative method. By the use of plasmids encoding hNaa15p (phNAA15-V5His) and hNaa16p (phNAA16-Flag), we could verify the specificity of the primers used. The relative amount of hNAA15/hNAA16 PCR in each sample was normalized to that of RpLP2 PCR and RPol2 independently.

Alignment and Tree Building

Homologues to human hNaa15p and hNaa16p were identified using them to search Ensembl version 38, April 2006 [35]. Ensembl peptide identifiers for hNaa15p and hNaa16p are ENSP00000296543 and ENSP000000368716/ENSP00000310683 respectively. The two identifiers for hNaa16p represent two possible alternative transcripts in the Ensembl genome annotation. Both hNaa15p and hNaa16p are members of Ensembl protein family ENSF00000002142. All the peptide sequences from this family were aligned using Muscle [36] with the default settings. Coding sequence alignments were produced by aligning the coding sequences with reference to the alignment of the corresponding peptide sequences. A tree was built from the coding sequence alignment using MrBayes [37] with different rates for transitions and transversions, and running for 250,000 generations. The first 100,000 generations were discarded as burn-in, after which the likelihood scores had converged. A consensus tree was built from the remaining 150,000 generations by sampling every 100th generation, and summarising as a majority-rule consensus tree. The resultant unrooted tree was rooted by treating Saccharomyces cerevisiae NAA15 (identifier: YDL040C) as an outgroup. Figure 6 was created with iTOL [38].

List of Abbreviations

ARD1: Arrest defective 1; hNatA: human N-terminal acetyltransferase A; LC: high performance liquid capillary; MS: mass spectrometry; Naa: N alpha acetyltransferase; NAT: N-terminal acetyltransferase.

Authors' Contributions

TA planned the study, performed the large scale immunoprecipitation experiments and wrote the manuscript draft. DG performed knockdown experiments, phenotype studies and quantitative RT-PCR. DK performed immunoprecipitation and acetyltransferase assays. MB performed the evolutionary analysis and alignments. KKS performed polysome experiments. DA performed the mass spectrometry analysis and peptide data analysis. All authors took part in planning and manuscript preparation. All authors read and approved the final manuscript.

Acknowledgements

We thank N. Glomnes and E. Skjelvik for technical assistance. This work was supported by The Norwegian Cancer Society (Grant to JEV), The Locus of Experimental Cancer Research, University of Bergen (Grant to JRL), The Meltzer Foundation (Grant to TA), FUGE and Western Norway Regional Health Authority (Grants to TA and JEV).

References

1. Arnesen T, Van Damme P, Polevoda B, Helsens K, Evjenth R, Colaert N, Varhaug JE, Vandekerckhove J, Lillehaug JR, Sherman F, Gevaert K: Proteomics analyses reveal the evolutionary conservation and divergence of N-terminal acetyltransferases from yeast and humans. Proc Natl Acad Sci USA 2009, 106:8157–8162.

2. Brown JL, Roberts WK: Evidence that approximately eighty per cent of the soluble proteins from Ehrlich ascites cells are Nalpha-acetylated. J Biol Chem 1976, 251:1009–1014.

3. Brown JL: A comparison of the turnover of alpha-N-acetylated and nonacetylated mouse L-cell proteins. J Biol Chem 1979, 254:1447–1449.

4. Jornvall H: Acetylation of Protein N-terminal amino groups structural observations on alpha-amino acetylated proteins. J Theor Biol 1975, 55:1–12.

5. Persson B, Flinta C, von Heijne G, Jornvall H: Structures of N-terminally acetylated proteins. Eur J Biochem 1985, 152:523–527.

6. Polevoda B, Sherman F: N-terminal acetyltransferases and sequence requirements for N-terminal acetylation of eukaryotic proteins. J Mol Biol 2003, 325:595–622.

7. Gautschi M, Just S, Mun A, Ross S, Rucknagel P, Dubaquie Y, Ehrenhofer-Murray A, Rospert S: The yeast N(alpha)-acetyltransferase NatA is quantitatively anchored to the ribosome and interacts with nascent polypeptides. Mol Cell Biol 2003, 23:7403–7414.

8. Park EC, Szostak JW: ARD1 and NAT1 proteins form a complex that has N-terminal acetyltransferase activity. EMBO J 1992, 11:2087–2093.

9. Mullen JR, Kayne PS, Moerschell RP, Tsunasawa S, Gribskov M, Colavito-She-panski M, Grunstein M, Sherman F, Sternglanz R: Identification and charac-terization of genes and mutants for an N-terminal acetyltransferase from yeast. EMBO J 1989, 8:2067–2075.

10. Whiteway M, Szostak JW: The ARD1 gene of yeast functions in the switch be-tween the mitotic cell cycle and alternative developmental pathways. Cell 1985, 43:483–492.

11. Whiteway M, Freedman R, Van Arsdell S, Szostak JW, Thorner J: The yeast ARD1 gene product is required for repression of cryptic mating-type informa-tion at the HML locus. Mol Cell Biol 1987, 7:3713–3722.

12. Ametzazurra A, Larrea E, Civeira MP, Prieto J, Aldabe R: Implication of human N-alpha-acetyltransferase 5 in cellular proliferation and carcinogenesis. Onco-gene 2008, 27:7296–7306.

13. Arnesen T, Anderson D, Baldersheim C, Lanotte M, Varhaug JE, Lillehaug JR: Identification and characterization of the human ARD1-NATH protein acetyl-transferase complex. Biochem J 2005, 386:433–443.

14. Starheim KK, Arnesen T, Gromyko D, Ryningen A, Varhaug JE, Lillehaug JR: Identification of the human N(alpha)-acetyltransferase complex B (hNatB): a complex important for cell-cycle progression. Biochem J 2008, 415:325–331.

15. Starheim KK, Gromyko D, Evjenth R, Ryningen A, Varhaug JE, Lillehaug JR, Arnesen T: Knockdown of the Human N{alpha}-Terminal Acetyltransferase Complex C (hNatC) Leads to p53-Dependent Apoptosis and Aberrant hArl8b Localization. Mol Cell Biol, in press.

16. Arnesen T, Anderson D, Torsvik J, Halseth HB, Varhaug JE, Lillehaug JR: Clon-ing and characterization of hNAT5/hSAN: an evolutionarily conserved com-ponent of the NatA protein N-alpha-acetyltransferase complex. Gene 2006, 371:291–295.

17. Arnesen T, Betts MJ, Pendino F, Liberles DA, Anderson D, Caro J, Kong X, Varhaug JE, Lillehaug JR: Characterization of hARD2, a processed hARD1 gene duplicate, encoding a human protein N-alpha-acetyltransferase. BMC Biochem 2006, 7:13.

18. Arnesen T, Gromyko D, Pendino F, Ryningen A, Varhaug JE, Lillehaug JR: In-duction of apoptosis in human cells by RNAi-mediated knockdown of hARD1 and NATH, components of the protein N-alpha-acetyltransferase complex. Oncogene 2006, 25:4350–4360.

19. Fisher TS, Etages SD, Hayes L, Crimin K, Li B: Analysis of ARD1 function in hypoxia response using retroviral RNA interference. J Biol Chem 2005, 280:17749–17757.

20. Lim JH, Park JW, Chun YS: Human arrest defective 1 acetylates and activates beta-catenin, promoting lung cancer cell proliferation. Cancer Res 2006, 66:10677–10682.

21. Arnesen T, Thompson PR, Varhaug JE, Lillehaug JR: The Protein Acetyltransferase ARD1: A Novel Cancer Drug Target? Curr Cancer Drug Targets 2008, 8:545–553.

22. Yi CH, Sogah DK, Boyce M, Degterev A, Christofferson DE, Yuan J: A genome-wide RNAi screen reveals multiple regulators of caspase activation. J Cell Biol 2007, 179:619–626.

23. Tatusova TA, Madden TL: BLAST 2 Sequences, a new tool for comparing protein and nucleotide sequences. FEMS Microbiol Lett 1999, 174:247–250.

24. Schultz J, Milpetz F, Bork P, Ponting CP: SMART, a simple modular architecture research tool: identification of signaling domains. Proc Natl Acad Sci USA 1998, 95:5857–5864.

25. Rappsilber J, Ryder U, Lamond AI, Mann M: Large-scale proteomic analysis of the human spliceosome. Genome Res 2002, 12:1231–1245.

26. Arnesen T, Gromyko D, Horvli O, Fluge O, Lillehaug J, Varhaug JE: Expression of N-acetyl transferase human and human Arrest defective 1 proteins in thyroid neoplasms. Thyroid 2005, 15:1131–1136.

27. Fluge O, Bruland O, Akslen LA, Varhaug JE, Lillehaug JR: NATH, a novel gene overexpressed in papillary thyroid carcinomas. Oncogene 2002, 21:5056–5068.

28. Paradis H, Islam T, Tucker S, Tao L, Koubi S, Gendron RL: Tubedown associates with cortactin and controls permeability of retinal endothelial cells to albumin. J Cell Sci 2008, 121:1965–1972.

29. Sugiura N, Adams SM, Corriveau RA: An evolutionarily conserved N-terminal acetyltransferase complex associated with neuronal development. J Biol Chem 2003, 278:40113–40120.

30. Eng JK, Mccormack AL, Yates JR: An Approach to Correlate Tandem Mass-Spectral Data of Peptides with Amino-Acid-Sequences in A Protein Database. Journal of the American Society for Mass Spectrometry 1994, 5:976–989.

31. Gururaja T, Li WQ, Bernstein J, Payan DG, Anderson DC: Use of MEDUSA-based data analysis and capillary HPLC-ion-trap mass spectrometry to examine complex immunoaffinity extracts of RbAp48. Journal of Proteome Research 2002, 1:253–261.

32. Anderson DC, Li WQ, Payan DG, Noble WS: A new algorithm for the evaluation of shotgun peptide sequencing in proteomics: Support vector machine

classification of peptide MS/MS spectra and SEQUEST scores. Journal of Proteome Research 2003, 2:137–146.

33. Pfund C, Lopez-Hoyo N, Ziegelhoffer T, Schilke BA, Lopez-Buesa P, Walter WA, Wiedmann M, Craig EA: The molecular chaperone Ssb from Saccharomyces cerevisiae is a component of the ribosome-nascent chain complex. EMBO J 1998, 17:3981–3989.

34. Vedeler A, Pryme IF, Hesketh JE: The characterization of free, cytoskeletal and membrane-bound polysomes in Krebs II ascites and 3T3 cells. Mol Cell Biochem 1991, 100:183–193.

35. Birney E, Andrews D, Caccamo M, Chen Y, Clarke L, Coates G, Cox T, Cunningham F, Curwen V, Cutts T, Down T, Durbin R, Fernandez-Suarez XM, Flicek P, Gräf S, Hammond M, Herrero J, Howe K, Iyer V, Jekosch K, Kähäri A, Kasprzyk A, Keefe D, Kokocinski F, Kulesha E, London D, Longden I, Melsopp C, Meidl P, Overduin B, Parker A, Proctor G, Prlic A, Rae M, Rios D, Redmond S, Schuster M, Sealy I, Searle S, Severin J, Slater G, Smedley D, Smith J, Stabenau A, Stalker J, Trevanion S, Ureta-Vidal A, Vogel J, White S, Woodwark C, Hubbard TJ: Ensembl 2006. Nucleic Acids Res 2006, 34:D556–D561.

36. Edgar RC: MUSCLE: multiple sequence alignment with high accuracy and high throughput. Nucleic Acids Res 2004, 32:1792–1797.

37. Ronquist F, Huelsenbeck JP: MrBayes 3: Bayesian phylogenetic inference under mixed models. Bioinformatics 2003, 19:1572–1574.

38. Letunic I, Bork P: Interactive Tree Of Life (iTOL): an online tool for phylogenetic tree display and annotation. Bioinformatics 2007, 23:127–128.

Biochemical Characterization and Cellular Imaging of a Novel, Membrane Permeable Fluorescent Camp Analog

Daniela Moll, Anke Prinz, Cornelia M. Brendel, Marco Berrera, Katrin Guske, Manuela Zaccolo, Hans-Gottfried Genieser and Friedrich W. Herberg

ABSTRACT

Background

A novel fluorescent cAMP analog (8-[Pharos-575]- adenosine-3', 5'-cyclic monophosphate) was characterized with respect to its spectral properties, its ability to bind to and activate three main isoenzymes of the cAMP-dependent protein kinase (PKA-Iα, PKA-IIα, PKA-IIβ) in vitro, its stability towards phosphodiesterase and its ability to permeate into cultured eukaryotic cells using resonance energy transfer based indicators, and conventional fluorescence imaging.

Results

The Pharos fluorophore is characterized by a Stokes shift of 42 nm with an absorption maximum at 575 nm and the emission peaking at 617 nm. The quantum yield is 30%. Incubation of the compound to RIIα and RIIβ subunits increases the amplitude of excitation and absorption maxima significantly; no major change was observed with RIα. In vitro binding of the compound to RIα subunit and activation of the PKA-Iα holoenzyme was essentially equivalent to cAMP; RII subunits bound the fluorescent analog up to ten times less efficiently, resulting in about two times reduced apparent activation constants of the holoenzymes compared to cAMP. The cellular uptake of the fluorescent analog was investigated by cAMP indicators. It was estimated that about 7 μM of the fluorescent cAMP analog is available to the indicator after one hour of incubation and that about 600 μM of the compound had to be added to intact cells to half-maximally dissociate a PKA type IIα sensor.

Conclusion

The novel analog combines good membrane permeability- comparable to 8-Br-cAMP – with superior spectral properties of a modern, red-shifted fluorophore. GFP-tagged regulatory subunits of PKA and the analog co-localized. Furthermore, it is a potent, PDE-resistant activator of PKA-I and -II, suitable for in vitro applications and spatial distribution evaluations in living cells.

Background

Fluorescent nucleotides have become widely utilized tools in basic research [1], and the number of corresponding reports on their use in cellular systems is vast. However studies involving the second messenger cAMP have not kept pace, and studies using fluorescently tagged cAMP analogs are still limited.

Early reports mainly used nucleobase-modified analogs where the purine ring system was part of the fluorophore, such as 1, N6- etheno-cAMP [2], 2- aza- 1, N6- etheno-cAMP [3] or the cyclic phosphate of 2- aminopurine riboside [4]. However, these compounds are far from being optimal with regard to membrane permeability, cAMP-dependent protein kinase (PKA) binding affinity and stability towards phosphodiesterases (PDEs). Further more, the fluorophores lack brilliance and possess unfavorable spectral properties, e.g., excitation is to be performed in the UV range, which can be harmful to intact cells and monitoring of the relatively short emission wavelengths is often disturbed by intrinsic fluorescent components of the cell.

For studies involving PDEs, anthraniloyl [5]- and methylanthraniloyl- modified [6] cAMP (MANT-cAMP) have been introduced, where the ribose 2- position carries a fluorescent reporter group. Since the 2'- modification render these structures unable to activate PKA, the MANT group has been linked to the positions 6 [7] and 8 [8] of the adenine nucleobase as well (MABA-cAMP). According to corresponding lipophilicity data (log kw), cyclic nucleotides with MANT modification have improved membrane permeability and better PDE-resistance (at least 6- and 8- modified structures), but are still suboptimal with respect to their spectral properties. The same holds true for cAMP modified with the NBD fluorophore (8-[2-[(7-Nitro-4-benzofurazanyl)amino]ethyl]thio]adenosine-3',5'-cyclic monophosphate; 8-NBD-cAMP) [6].

Fluorescein and rhodamine have been attached to the 8- position of cAMP as well [9], however, in spite of the improved spectral properties of these dyes, both conjugates are not membrane permeable due to an additional charge within the dye moiety. Unfortunately, nearly all modern fluorescent dye structures contain positive or negative charges which support the electronic push/pull mechanism of the respective chromophore but render the corresponding conjugates rather polar, especially when attached to nucleotides with their polar phosphate groups. Even cAMP conjugates with state-of-the-art dyes such as Cy3 [10], Evoblue, and Bodipy® [11,12], which have excellent spectral properties, fail to pass cellular membranes, and require invasive application techniques like patch clamp or microinjection or the osmotic lysis of pinocytic vesicles [13] and are mainly utilized in vitro assays. Importantly, if these dyes are connected to the 2'- ribose moiety, the resulting conjugates will not bind to and activate PKA anymore. Finally, phosphate-modified caged cAMP analogs have been described, which – upon photo-activation – release cAMP together with fluorescent coumarines [14,15].

Thus, in spite of quite a number of different fluorescent variants of cAMP, these structures have only limited application scopes, and presently no analog is available that offers improved properties in all important aspects mentioned, and which could be used with intact cells for tracking or intracellular imaging experiments.

The main effector enzyme of cAMP is the cAMP-dependent protein kinase (PKA), which reversibly phosphorylates substrate proteins. Protein kinases and their counter players, phosphatases and cAMP-degrading PDEs, are key regulatory enzymes in eukaryotic cells. PKA is a multi-substrate enzyme mediating the majority of the known effects of cAMP by regulating the activity of proteins involved in signal transduction, energy metabolism, cell proliferation, and differentiation [16]. In the absence of cAMP, all PKA isoforms consist of two regulatory (R) and two catalytic subunits (C) that form an inactive tetramer. Binding of at total of four cAMP molecules to the two tandem cAMP binding sites of each R subunit promotes the dissociation of the holoenzyme complex and leads to the release of the now active C subunits phosphorylating target proteins in the cytosol or in the nucleus [17]. Main isoforms of the C subunit are Cα, Cβ, Cγ and PrKX, and several minor isoforms have been identified at least at cDNA level. In human, four different isoforms of the R subunit (RIα, RIβ, RIIα, RIIβ) have been identified [18].

Besides PKA and its corresponding signaling pathway, cAMP addresses additional cellular targets such as cyclic nucleotide-dependent ion channels (cyclic nucleotide gated ion channel and hyperpolarization-activated cyclic nucleotide-modulated channel) as well as the exchange protein directly activated by cAMP (Epac), which are worthwhile objects for evaluation with fluorescent analogs [19,20].

In view of the importance of the cAMP messenger system, improved tools for a more detailed investigation of functions and receptor distribution would be quite helpful. Thus, in this study, the properties of a novel commercially available conjugate of cAMP with the Pharos dye, 8-[Pharos-575]- adenosine-3', 5'-cyclic monophosphate (8-[φ-575]-cAMP) and the corresponding free dye were investigated and analyzed physico-chemically (spectral properties, stability), with respect to PKA-RI and -RII subunit binding as well as holoenzyme activation in vitro and in living cells, PDE-resistance, and cellular uptake.

Results and Discussion

Photochemical Characterization of the Pharos Dye

First we determined the photo-chemical properties of the free Pharos dye. The Stokes shift at pH 6.0, pH 7.0 and pH 7.4 was 42 nm resulting from an absorption maximum at 575 nm and an emission maximum at 617 nm (Fig. 1; ε_{575nm} = 15,650). 8-[φ-575]-cAMP behaved similarly with an excitation maximum at 577 nm and emission maximum at 605 nm at pH 7.0 (Fig. 2, and data not shown).

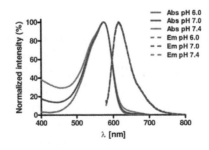

Figure 1. Fluorescence absorption and emission spectra of the Pharos dye. The absorption maxima of the Pharos dye dissolved in buffer with the indicated pH values is at 575 nm, whereas the emission maxima are at 617 nm. All spectra exhibit a large Stokes shift of 42 nm.

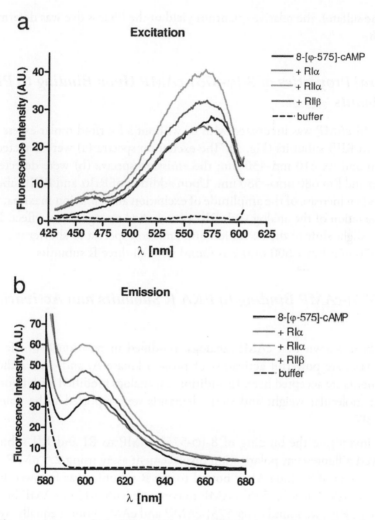

Figure 2. Fluorescence spectra of 8-[φ-575]-cAMP bound to PKA R subunits. 150 nM 8-[φ-575]-cAMP was incubated with or without fourfold molar excess of PKA R subunits as indicated in the figure. The excitation spectra (a) were detected at Em 617 nm and Ex 610 nm-430 nm; the emission spectra (b) were detected at Ex 575 nm and Em 680 nm-580 nm. The experiments were repeated four times with similar results.

To calculate the relative quantum yield φ of the Pharos dye the absorption (500 nm to 640 nm) and fluorescence (λex = 553 nm, λem ranging from 558 nm to 800 nm) spectra of several Pharos dye concentrations (640 nM to 64.2 μM) were recorded using quinine sulfate as a standard [21,22]. The determination of the relative quantum yield is generally accomplished by plotting the integrated fluorescence intensity versus the absorbance at the excitation wavelength. By comparing the slope of the linear regression with that of a standard substance (quinine sulfate), the relative quantum yield of the Pharos dye was determined to be 30%.

Spectral Properties of 8-[φ-575]-cAMP Upon Binding to PKA R-Subunits

8-[φ-575]-cAMP was incubated with or without a fourfold molar excess of RIα, RIIα and RIIβ subunits (Fig. 2). The excitation spectra (a) were detected at Em 617 nm and Ex 610 nm-430 nm; the emission spectra (b) were detected at Ex 575 nm and Em 680 nm – 580 nm. Upon addition of RIIα and RIIβ subunits we observed an increase of the amplitude of excitation and emission maxima, whereas the interaction of the analog with RIα protein had no significant effect. Furthermore, a slight shift of the excitation and emission spectra to shorter wavelengths (Ex = 570 nm; Em = 600 nm) was found with all three R subunits.

8-[φ-575]-cAMP Binding to PKA R Subunits and Activation of PKA

It has been shown that cAMP analogs, modified in position 8 of the adenine nucleobase, are powerful activators of protein kinase A, and even rather bulky substituents are accepted here. In addition, all analogs modified with fluorophores of high molecular weight and steric demands were reported to be potent PKA agonists[7].

To investigate the binding of 8-[φ-575]-cAMP to RI and RII subunits, we employed a fluorescent polarization displacement assay using 2.5 nM R subunit. Displacement of 8-Fluo-cAMP bound to the R subunits was followed by allowing either cAMP or 8-[φ-575]-cAMP to compete with 8-Fluo-cAMP binding. In the case of RIα, we found 8-[φ-575]-cAMP and cAMP bound equally well to the R subunit (Fig. 3a). In case of RIIα and RIIβ, cAMP was about 10 times and 5 times more efficient in displacing 8-Fluo-cAMP compared to 8-[φ-575]-cAMP, respectively (Fig. 3b und 3c). With respect to both, isoform and site selectivity, and considering the bulky substituent in position 8 of the adenine nucleobase, it could have been expected that 8-[φ-575]-cAMP prefers the B- site of PKA type II. However, our data show that binding to RIα is comparable to cAMP, whereas a lower affinity of 8-[φ-575]-cAMP for RII isoforms was detected. In this respect, 8-[φ-575]-cAMP acts similar to 8-Fluo-cAMP, which shows rather high affinity for the site BII along with rather lower affinity to AII, but a quite equal binding capacity to both sites A and B of RI, thus resulting in an overall higher binding to RIα [23], analogous to other 8-substituted analogs [24,25].

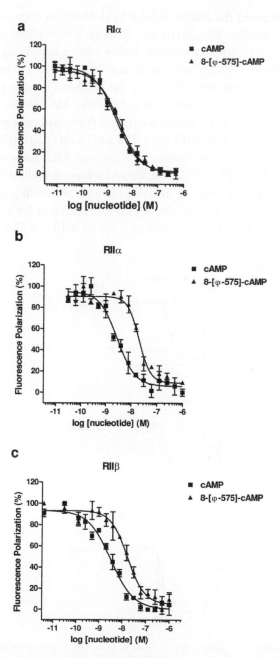

Figure 3. Competitive nucleotide binding assay using fluorescence polarization. Serial dilutions of cAMP or 8-[φ-575]-cAMP in the presence of 1 nM 8-Fluo-cAMP were prepared. 2.5 nM R subunit was added and fluorescence polarization was determined at Ex485 nm/Em535 nm. EC50 values for cAMP and 8-[φ-575]-cAMP binding to the R subunit isoforms were deducted from the corresponding titration curves. Each data point represents the mean +/- S.E.M. from at least triplicate measurements.

We next investigated the ability of 8-[φ-575]-cAMP to activate PKA-I and -II holoenzymes side by side. RIα, RIIα and RIIβ each were allowed to form a holoenzyme complex with the Cα subunit, before adding increasing amounts of either cAMP or 8-[φ-575]-cAMP to re-activate the Cα-subunit. Kinase activity was assayed spectrophotometrically [26], using the synthetic heptapeptide Kemptide as a substrate. The binding properties of 8-[φ-575]-cAMP to the RIα subunit is reflected in a nearly identical activation titration curve and corresponding activation constant comparing analog (EC50 = 150 nM) with cAMP (EC50 = 120 nM, [27,28], Fig. 4a). However, activation of RIIα and RIIβ holoenzymes was less cooperative using 8-[φ-575]-cAMP, indicated by a more shallow hill slope of the activation titration curves (Fig. 4a and 4b). The corresponding apparent activation constants (EC50-values) were about two-fold increased for 8-[φ-575]-cAMP compared to cAMP (EC50 RIIα = 280 nM; EC50 RIIβ = 900 nM).

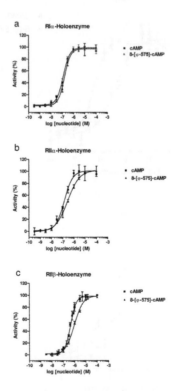

Figure 4. 8-[φ-575]-cAMP is a potent PKA activator in vitro. For determination of apparent activation constants, purified recombinant R subunits were allowed to form holoenzyme complexes with PKA-Cα (20 nM) as detailed in the methods section. Activation assays were performed by increasing (0.3 nM-10 μM) cAMP. To obtain apparent activation constants (Kact), the normalized activity of PKA-Cα was plotted against the logarithm of the cAMP (■) and 8-[φ-575]-cAMP (▲) concentration and fitted according to a sigmoid dose-response model (Graphpad Prism, variable slope). Each data point represents the mean ± S.D. of two measurements. Experiments were repeated two to three times with similar results.

Uptake of 8-[φ-575]-cAMP in Living Cells and Activation of PKA

Insights into the intracellular distribution within eukaryotic cells were obtained by intracellular imaging of the cyclic nucleotide analog. 8-[φ-575]-cAMP clearly enters the cultivated HEK293 and CHO cells. In most cases it displays a spotty distribution inside the cells and seems to accumulate in bright aggregates over a diffusely labeled background. The nuclear compartment was unlabeled in most cells (Fig. 5a and 5b). In some experiments, employing COS-7 cells less than 5% of the cells showed an accumulation of the compound in the nucleus, which might be attributable to apoptotic degeneration of the cells (data not shown). Variations in the incubation procedure (e.g. decrease in temperature, addition of pluronic® or serum during incubation [29,30]) did not significantly change the distribution pattern or the accumulation in bright spots. However, an incubation temperature of 4°C led to a markedly reduced uptake and accumulation of the compound (data not shown). This could indicate that membrane trafficking, which is inhibited at 4°C, is part of the accumulation process. Whether pinocytotic uptake of the compound or exocytotic processes involving cAMP are inhibited [31], deserves further study. The free Pharos dye efficiently diffuses within the cells and does not accumulate in spots like 8-[φ-575]-cAMP does (Fig. 5c and 5d).

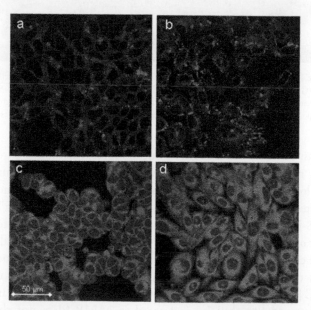

Figure 5. Visualization of compounds in living cells. Intracellular imaging of 8-[φ-575]-cAMP (a-b) and of Pharos dye (c-d) in HEK293 (a,c) and in CHO (b,d) cells after 1 hour of treatment.

It has previously been shown that fluorescent cAMP analogs were able to label RIα aggregates, respectively [32-34]. We therefore investigated whether the compound co-localizes with GFP-tagged RIα and RIIα proteins expressed in COS-7 cells, and indeed we found co-localization of 8-[φ-575]-cAMP, but not the free dye, with R subunits (Fig. 6, and data not shown), indicating that the accumulation of the fluorescent cAMP might be in part due to association to (clustered) cAMP binding proteins. However, we can not entirely exclude unspecific aggregation of 8-[φ-575]-cAMP.

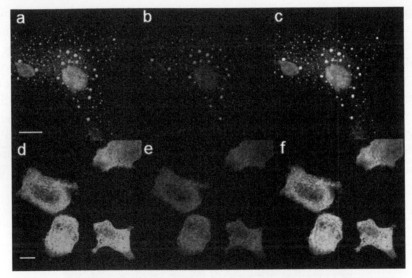

Figure 6. Co-localization of 8-[φ-575]-cAMP and GFP-tagged R subunits. COS-7 cells were transiently transfected with GFP-hRIα (a-c) and GFP-hRIIα (d-f), incubated with 500 µM 8-[φ-575]-cAMP for 30 minutes. Cells were fixed and fluorescence was imaged using confocal microscopy: (a,c) green fluorescence of GFP (b,e), red fluorescence of 8-[φ-575]-cAMP, (c,f) merged images. The scale bar indicates 10 µm.

In contrast to many other modern fluorophores, the Pharos chromophore has considerably reduced bulkiness along with high lipophilicity. Thus, 8-[φ-575]-cAMP has a lower molecular weight compared to e.g. fluorescein-modified-cAMP (8-[2-[(Fluoresceinylthioureido) amino]ethyl]thio]adenosine-3', 5'-cyclic monophosphate; 8-Fluo-cAMP). Due to its merely hydrophobic character, the dye has a big impact on the analog overall lipophilicity, which in turn should compensate the negative charge of the cyclic phosphate and finally lead to good membrane permeability of 8-[φ-575]-cAMP. In lipophilicity measurements using reversed phase HPLC [35], its logKw was determined to be 2.95 resulting in a more than 70 times increased lipophilicity compared to cAMP (data not shown). In this respect the analog resembles highly membrane-permeant PKA agonists such as

Sp-5,6-DCl-cBIMPS [36] or the Epac activator 8-pCPT-2'-O-Me-cAMP [37], and surpasses all fluorescent cAMP analogs described so far. We therefore investigated the kinetics of 8-[φ-575]-cAMP uptake in living cells using a genetically encoded fluorescent indicator (H30) formed by a cAMP binding domain from Epac and two spectral variants of the green fluorescent protein (GFP) [38]. Its functioning is based on the phenomenon of fluorescence resonance energy transfer (FRET). FRET relies on a non-radiative, distance-dependent transfer of energy between the two fluorescent domains: by exciting the first, the emission from the second one can be detected. Since FRET depends on the distance between the two fluorophores which in turn is ligand-dependent, this probe is used to estimate the intracellular level of cAMP or -analogs.

The cells were treated with 500 μM 8-[φ-575]-cAMP for one hour and, during this period, FRET ratio increased linearly (Fig. 7a–d). On the basis of the cAMP dose-FRET response curve [39], we estimate that about 7 μM 8-[φ-575]-cAMP is available to the intracellular fluorescent indicator. The same experiments were performed using Sp-5,6-DCl-cBIMPS, a highly membrane permeable cAMP analog [36]. In this case, after 40 minutes of treatment, the fluorescent probe was almost completely saturated (Fig. 7e–h), indicating that Sp-5,6-DCl-cBIMPS crosses the plasma membrane more efficiently compared to 8-[φ-575]-cAMP, in contrast to the retention based lipophilicity measurements on HPLC, where the analogs behaved more similarly. We can not exclude, however, that the Epac- based FRET sensor used for the measurements preferentially binds Sp-5,6-DCl-cBIMPS.

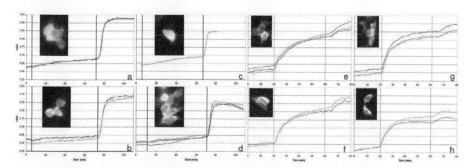

Figure 7. Uptake of 8-[φ-575]-cAMP in intact cells. Kinetics of 8-[φ-575]-cAMP (a-d) and of Sp-5,6-DCl-cBIMPS (e-h) uptake in living cells. HEK293 cells were transfected for transient expression of the H30 indicator of cAMP and treated with 100 μM IBMX to inhibit phosphodiesterases. The ratio between the background subtracted emission intensities at 480 nm and 545 nm is plotted as a function of time. The vertical bars indicate the administrations of either 500 μM 8-[φ-575]-cAMP (a-d) or Sp-5,6-DCl-cBIMPS (e-h) and of 25 μM Forskolin, which activates adenylyl cyclase and saturates the FRET-based probe. The insets show the yellow fluorescent protein fluorescence at the beginning of the experiments and the regions of interest where FRET ratios are calculated.

As the reduced intracellular availability of 8-[φ-575]-cAMP could also be due to intracellular PDE-mediated degradation, we tested stability of the compound towards a cAMP-specific PDE isoform (PDE4D5) in vitro using a coupled spectrophotometric activity assay [40]. No degradation of 8-[φ-575]-cAMP was detected during the 15 minutes of assay duration (data not shown). The specific activity of the enzyme was 0.6 U/mg for cAMP. It has been demonstrated before, that 8-modified cAMP analogs possess considerably high resistance towards PDE degradation [41]. Absolutely stable analogs would arise from conjugating the Pharos fluorophore to phosphorothioate- modified structures such as Rp- or Sp-cAMPS (Rp- or Sp-diastereomer of adenosine 3'-5'-monophosphorothiorate).

We next tested if 8-[φ-575]-cAMP can activate the PKA holoenzyme in living COS-7 cells using a recently established bioluminescence resonance energy transfer (BRET) based reporter for PKA-IIα, based on transient co-expressed luciferase-tagged RIIα and (RIIα-RLuc) as the BRET donor and a GFP-tagged Cα subunit (GFP2-Cα) as the BRET acceptor. In this assay, a decrease in BRET signal indicates intracellular dissociation of the PKA holoenzyme [42]. As depicted in figure 8a, the cAMP analog efficiently allows the PKA subunits to dissociate. Half-maximal holoenzyme dissociation was achieved at a concentration of about 600 μM analog added to the cells. This value in the same range as the value for e.g. 8-Br-cAMP (EC50 = 1.5 mM [42]) but is surpassed by acetoxymethyl esters of cyclic nucleotides, that activate PKA in the low μM range determined by BRET [7] as well as in physiological assays [43]. Finally, we performed a bystander BRET test [44], where COS-7 cells were transfected with a constant amount of donor-expression plasmid (RIIα-RLuc, 0.5 μg) and with increasing amounts of acceptor-expression plasmid (GFP2-Cα, 0–2 μg). The cells were incubated with 600 μM 8-[φ-575]-cAMP for 30 minutes prior to the BRET ratio determination, or mock treated. Figure 8b depicts the normalized BRET-values of two experiments each performed with n = 6 wells per incubation. When no BRET acceptor is expressed in the cells, we observed the background BRET value, and the incubation with 8-[φ-575]-cAMP has no effect. With increasing amounts of acceptor DNA, the BRET values rises and reaches an asymptote at the 1:1 molar ratio at 0.5 μg donor and acceptor coding DNA each, as expected from a 1:1 interaction of PKA subunits in the holoenzyme. Thereafter the BRET value does not increase further, indicating a specific interaction, which can be prohibited to about 50% by incubation with 600 μM 8-[φ-575]-cAMP, as determined previously (see figure 8a and 8b).

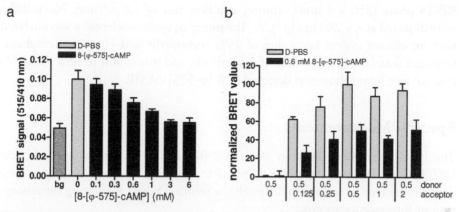

Figure 8. 8-[φ-575]-cAMP activates intracellular PKA. (a) For standard BRET experiments, COS-7-cells were co-transfected with the PKA type II sensor construct or negative control plasmids (bg) as indicated and grown for 48 hours. Cells were treated with the indicated amount of 8-[φ-575]-cAMP for 30 minutes, or mock treated (D-PBS). BRET signals were obtained after addition of the luciferase substrate DeepBlueC™ and detection of luciferase and fluorescence light emission using a multi-label reader. Shown is a representative experiment, repeated three times; data are mean ± S.E.M., performed with n = 6 replicates. (b) A BRET titration experiment was performed as described in the methods section. Briefly, cells were co-transfected with a constant amount of BRET donor (hRIIα-Rluc) and an increasing amount of acceptor DNA (GFP2-hCα) as indicated. Before BRET read-out, cells were incubated with 0.6 mM 8-[φ-575] cAMP as described above. The BRET values of two independent experiments, each performed with n = 6 replicates, were background subtracted, normalized and plotted as mean ± S.E.M.

Conclusion

We can conclude that 8-[φ-575]-cAMP with its modern, red-shifted fluorophore is a useful, stable tool for in vitro and in vivo investigation of cAMP binding proteins. It spontaneously enters eukaryotic cells and can be sequestered to PKA regulatory subunits. However, undesired partial accumulation in the cell could not be entirely excluded in this study. It efficiently binds to and activates PKA-I and -II, and should have great potential for many cell biological and in vitro applications.

Methods

Materials

All cyclic nucleotides, as well as the free Pharos chromophore were derived from Biolog Life Science Institute (Bremen, Germany). The solubility of the novel Pharos compound is approximately 50 mM in water. LogKw data were determined by a retention-based lipophilicity ranking using a LiChrograph HPLC (Merck-Hitachi, Darmstadt, Germany) equipped with a reversed phase YMC

RP-18 phase (250 × 4 mm) running at a flow rate of 1.0 ml/min. Nucleotides were detected at λ = 280 nm [35]. A. The purity of cyclic nucleotides was analyzed with an elution system consisting of 25% acetonitrile and 10 mM triethyl ammonium formiate at a flow rate of 1.5 ml/min and was found to be > 99%. No trace of free fluorophore was detected in 8-[φ-575]-cAMP.

Spectral Measurements

The Pharos dye was diluted in 20 mM 2-(4-morpholino)-ethane sulfonic acid (MES) buffer adjusted to pH 6.0, pH 7.0 and pH 7.4, and all samples were degassed by bubbling nitrogen through the solution before measuring the absorption or fluorescence spectra.

For Stokes shift and quantum yield determination, absorption and fluorescence spectra were recorded in cells with 1 cm path length using a PerkinElmer Lambda 900 UV-Vis spectrophotometer and a Hitachi F-4500 fluorescence spectrophotometer, respectively.

To calculate the relative quantum yield φ of the free Pharos dye, the absorption (500 nm to 640 nm) and fluorescence (λex = 553 nm, λem ranging from 558 nm to 800 nm) spectra of several Pharos dye concentrations (640 nM to 64.2 µM) were recorded in A. bidest. The integrated fluorescence intensity (I) was plotted versus the absorbance (A) at the excitation wavelength. For low concentrations the dye molecules are not influenced by each other and thus exhibit a linear behavior concerning absorption and emission of light. Linear regression yields the slope $\Delta I/\Delta A$ that is compared to the corresponding data of a quantum yield standard by using the equation $\Box sa = (\Delta Isa/\Delta Asa)\cdot(\Delta Ast/\Delta Ist)\cdot\Box st$. As a standard, a solution of quinine sulfate in 1.0 N sulfuric acid ($\Box st$ = 0.546 with λex = 365 nm) was used [21,22].

For emission and excitation spectra (Kontron SFM25) with and without purified regulatory subunit of PKA (see below), 100 nM 8-[φ-575]-cAMP was measured with and without fourfold molar excess of regulatory subunit in 20 mM MOPS, 150 mM NaCl, 1 mM β-mercaptoethanol, pH (buffer A) at room temperature. The excitation spectra were detected at λem = 617 nm with λex ranging from 610 nm to 430 nm; emission spectra were measured with λex = 575 nm with λem ranging from 800 nm to 580 nm.

Expression and Purification of PKA Subunits

For expression of PKA regulatory subunits, one liter of Luria Broth medium containing 100 µg/ml of ampicillin was inoculated with E. coli BL21 (DE3)

Codon Plus RIL cells (Stratagene) transformed with human RIα, RIβ, RIIα (in pRSETB) or rat RIIβ (in pETIIc) and grown at 37°C to an OD600 nm of 0.8. Recombinant protein expression was induced by addition of 0.2 M isopropyl-β-D-thiogalactopyranoside (IPTG) and the culture was incubated at 25°C for additional 17–18 h. The pellets were stored at -20°C.

For purification of RI isoforms (Moll et al., submitted), cell lysis of protein expressing E. coli cells was performed two times with a French Pressure Cell (Thermo Electron Corp., Needham Heights, USA) in lysis buffer (20 mM 3-(N-morpholino) propane sulfonic acid (MOPS), 100 mM NaCl, 1 mM β-mercaptoethanol, 2 mM EDTA, 2 mM EGTA, pH7.0). The lysate was centrifuged at 27 000 ×g for 30 min and 4°C. 1.2 μmol Sp-8-AEA-cAMPS-agarose (Biolog Life Science Inst.) was used per purification, corresponding to 300–450 μl agarose-slurry. The protein content in 12 ml clarified supernatant was batch bound by gentle rotation over night by 4°C. The agarose was washed seven times with 1.25 ml lysis buffer. The protein elution step was performed with 1.25 ml of 10 mM cGMP in buffer A by gentle rotation at 4°C for 1 h. The agarose was rinsed with two additional wash steps (each 825 μl) with buffer A. Subsequently, the R subunits were subjected to gel filtration (PD10, Amersham Pharmacia Biotech, Freiburg, Germany) into buffer A. To remove all cGMP, the R subunits were dialyzed excessively against buffer A.

For purification of RII isoforms (Moll et al., submitted), cell lysis was performed in buffer consisting 20 mM MES, pH6.5, 100 mM NaCl, 5 mM EDTA, 5 mM EGTA and 5 mM β-mercaptoethanol with added protease inhibitors (PI): Leupeptin (0.025 mg/100 ml), TPCK and TLCK (each 1 mg/100 ml) (buffer B). after centrifugation at 27 000 ×g for 30 min and 4°C, the supernatant was precipitated at 4°C with 50% saturated ammonium sulfate (AS) for RIIα and 45% AS for RIIβ and centrifuged by 10 000 ×g, 15 min 4°C. The AS pellets were re-suspended in buffer B and protein was batch bound to 1.4 μmol settled Sp-8-AEA-cAMPS-agarose. The agarose was rinsed twice with 20 mM MOPS, pH7.0, 1 M NaCl, 5 mM β-mercaptoethanol and then two times with 10 ml buffer B. Subsequently, two elution steps were carried out with 25 mM cGMP in buffer B. RII subunits were subjected to gel filtration into 20 mM MES, pH 6.5,150 mM NaCl, 2 mM EDTA, 2 mM EGTA, 1 mM β-mercaptoethanol.

Murine Cα subunit (in pRSETB) was expressed in E. coli BL21 (DE3) (Stratagene) and purified as published previously [45,46]. Protein expression and purification was followed by SDS-polyacrylamid gel electrophoresis [47]. Typically, the recombinant proteins were purified to 95% homogeneity or higher.

Fluorescence Polarization (FP)

The fluorescence polarization displacement assay was performed as described before[7]. Increasing concentrations of cAMP or 8-[φ-575]-cAMP were mixed with 1 nM 8-Fluo-cAMP before adding 2.5 nM regulatory subunit RIα, RIIα, RIIβ. Fluorescence polarization was measured after 5 minutes of incubation at room temperature.

Determination of Activation Constants

PKA activity was assayed by the coupled spectrophotometric assay first described by Cook et al. [26] using 260 µM Kemptide (LRRASLG) as the substrate. Holoenzyme formation was carried out for 3 minutes at room temperature with 20 nM murine PKA-Cα subunit and an about 1.2 fold molar excess cAMP-free RIα, RIIα, or RIIβ subunit in assay mixture (10 mM $MgCl_2$, 100 µM ATP, 100 mM MOPS, 1 mM PEP, LDH, pyruvate kinase, NADH, 5 mM β-mercaptoethanol, pH 7.0). Apparent activation constants (K_{act}, EC_{50}) were determined by adding increasing amounts of cAMP or 8-[φ-575]-cAMP (0.3 nM to 10 µM).

BRET Assay

COS-7 cells were used for BRET experiments. They were routinely passaged and seeded in opaque 96-well microplates (CulturPlate™-96, PerkinElmer) 24 hours prior to co-transfection with the previously described PKA-IIα sensor, comprised of RIIα-RLuc (donor) and GFP2-Cα (acceptor) [42]. Two days following transfection with 0.5 µg donor and acceptor DNA, respectively, cells were rinsed with glucose-supplemented Dulbecco's PBS (D-PBS, Invitrogen), and subsequently incubated with 8-[φ-575]-cAMP (0.01–6 mM final concentration in D-PBS, prepared from a 20 mM stock solution), or mock treated for 30 minutes at room temperature. For the BRET read-out, the luciferase substrate DeepBlueC™ (PerkinElmer) was added at a final concentration of 5 µM in a total volume of 50 µl D-PBS. Light output was detected consecutively using a Fusion™ α-FP microplate reader (PerkinElmer, read time 1s, gain 25) equipped with appropriate filters for the donor (RLuc; λ = 410 nm ± 80 nm) and for the acceptor fluorophore (GFP²; λ = 515 nm ± 30 nm) emission. Emission values obtained with untransfected (n.t.) cells were routinely subtracted, and BRET signals were calculated as follows: ($emission_{(515nm)}$ − n.t. $cells_{(515nm)}$)/($emission_{(410nm)}$ − n.t. $cells_{(410nm)}$). Control measurements with cells expressing RLuc and GFP proteins without a fusion partner yield the background BRET signal. A BRET titration (bystander BRET test) was performed by co-transfection of COS-7 cells with a constant donor-expression plasmid (0.5 μg) with an increasing amount of acceptor-expression plasmid (0–2 μg). The cells were treated as described above. Prior to the BRET read-out, cells were incubated with or without 0.6 mM 8-[φ-575]-cAMP for 30 minutes.

Uptake of 8-[φ-575]-cAMP in Living Cells

HEK293 and CHO cells were grown in the Dulbecco's modified Eagle medium (DMEM, Invitrogen) and F12 nutrient mixture (HAM, Invitrogen) mediums, respectively, containing 10% FBS and supplemented with 2 mM L-glutamine, 100 U/ml penicillin, and 100 μg/ml streptomycin (all: Sigma-Aldrich), in a 37°C humidified atmosphere containing 5% CO_2. These cell lines are routinely used in our laboratory for FRET experiments using various sensors and they have been thoroughly characterized for fluorescent sensor expression levels.

For transient expression of the H30 sensor, cells were seeded onto 24-mm diameter round glass coverslips, and transfections were performed at 50–70% confluence with FuGENE-6 transfection reagent (Roche). Imaging experiments were performed after about 24 h. Cells were maintained in Hepes-buffered Ringer-modified solution, containing 125 mM NaCl, 5 mM KCl, 1 mM Na3PO4, 1 mM MgS04, 5.5 mM glucose, 1 mM CaCl2, and 20 mM Hepes, pH 7.4, at room temperature and treated with 100 μM 3-Isobutyl-1-methylxanthine (IBMX, Sigma-Aldrich) 10 minutes before the experiments. Cells were imaged on an inverted microscope (IX50; Olympus) with a 60× oil immersion objective (Olympus). The microscope was equipped with a monochromator (Polychrome IV; TILL Photonics) and a beam-splitter optical device (Multispec Microimager; Optical Insights). FRET variations were measured as changes of the ratio between the background-subtracted fluorescence emission intensities at 480 nm and 545 nm, on excitation at 430 nm. Forskolin (25 μM, Sigma-Aldrich) was added to saturate the FRET probe and to determine the maximal FRET response.

For the imaging of the intracellular distribution of 8-[φ-575]-cAMP and the Pharos dye, cells were grown and seeded as above. Before the image acquisitions, cells were treated with 100 μM IBMX and either 500 μM 8-[φ-575]-cAMP or the Pharos dye for 1 hour. Cells were then washed with the Hepes-buffered Ringer-modified solution described above and imaged on a confocal microscope (Leica) with a 20× oil immersion objective (Leica). Images were obtained by collecting the emission light from 600 nm to 640 nm, on excitation at 514 nm.

Intracellular co-localization of 8-[φ-575]-cAMP and GFP-hRIα or GFP-hRIIα was examined in COS-7 cells after two days transient expression of the GFP-tagged R subunits [42]. Cloning of the R subunits into the expression vector hpGFP2-N2 (PerkinElmer) was performed as described previously [48]. The cells were incubated for 30 minutes at 37°C with 500 µM 8-[φ-575]-cAMP in D-PBS, rinsed tree times with D-PBS, followed by a standard fixation procedure [42]. Cellular imaging was performed on a confocal microscope (Leica) with a Plan apo 100× oil immersion objective (Leica).

Statistical Procedures

Measurement and statistical evaluation of FP, kinase activity and BRET assays was carried out using GraphPad Prism software version 4 (GraphPad Software).

Abbreviations

8-[φ-575]-cAMP: 8-[Pharos-575]adenosine-3',5'-cyclic monophosphate; 8-Fluo-cAMP: 8-[2-[(Fluoresceinylthioureido)amino]ethyl]thio]adenosine -3',5'-cyclic monophosphate; Sp-5,6-DCl-cBIMPS: 5,6-dichlorobenzimidazole riboside -3',5'-cyclic monophosphorothioate, Sp-isomer; 8-pCPT-2'-O-Me-cAMP: 8-(4-chlorophenylthio)-2'-O-methyladenosine -3',5'-cyclic monophosphate; 8-Br-cAMP: 8-Bromoadenosine -3',5'-cyclic monophosphate; Sp-8-AEA-cAMPS-agarose: 8-(2-aminoethylamino)adenosine -3',5'-cyclic monophosphorothioate, Sp- isomer, immobilized to agarose; MOPS: 3-(N-morpholino) propane sulfonic acid; MES: 2-(4-morpholino)-ethane sulfonic acid; PKA: protein kinase A, cAMP-dependent protein kinase; R: regulatory subunit; C: catalytic subunit; FRET: Fluorescence resonance energy transfer; BRET: bioluminescence resonance energy transfer; PDE: phosphodiesterase; IBMX: 3-Isobutyl-1-Methylxanthine.

Authors' Contributions

H–GG synthesized the Pharos compounds and performed studies on stability and lipophilicity, DM and CMB performed the physical characterization experiments, DM performed the FP assay, AP coordinated the study, performed BRET and kinase activity assays, cellular distribution evaluation and co-localization experiments; MB performed FRET assays and cellular distribution evaluation, KG performed cellular co-localization experiments and the PDE assay, FWH and MZ contributed to the design of the experiments in this study and the data evaluation, AP and H–GG wrote the manuscript.

Acknowledgements

We thank M. Diskar, C. Demme, M. Hantsch and H.-M. Zenn for technical assistance. H. Rühling (University of Kassel) is acknowledged for confocal microscopy imaging. Plasmids for expression of hRIα, hRIIα rRIIβ and mCα were a kind gift of Prof. Dr. S.S. Taylor, UC San Diego, U.S.A. Recombinant PDE4D5 enzyme was kindly provided by Prof. Dr. M.D. Houslay, University of Glasgow, Scotland. This project was supported by grants of the EU (LSHB-CT-2006-037189) to FWH, MZ, and HG, BMBF NGFN2 (FKZ01GR0441) and DFG (He 1818/4) to FWH, CMB is supported by a Ph.D. grant of the University of Kassel.

References

1. Cremo CR: Fluorescent nucleotides: synthesis and characterization. Methods Enzymol 2003, 360:128–177.

2. Builder SE, Beavo JA, Krebs EG: Stoichiometry of cAMP and 1,N6-etheno-cAMP binding to protein kinase. J Biol Chem 1980, 255(6):2350–2354.

3. Tsou KC, Yip KF, Lo KW: 1,N6-etheno-2-aza-adenosine 3',5'-monophosphate: a new fluorescent substrate for cycle nucleotide phosphodiesterase. Anal Biochem 1974, 60(1):163–169.

4. Scott SP, Tanaka JC: Molecular interactions of 3',5'-cyclic purine analogues with the binding site of retinal rod ion channels. Biochemistry 1995, 34(7):2338–2347.

5. Alfonso A, Estevez M, Louzao MC, Vieytes MR, Botana LM: Determination of phosphodiesterase activity in rat mast cells using the fluorescent cAMP analogue anthraniloyl cAMP. Cell Signal 1995, 7(5):513–518.

6. Hiratsuka T: New fluorescent analogs of cAMP and cGMP available as substrates for cyclic nucleotide phosphodiesterase. J Biol Chem 1982, 257(22):13354–13358.

7. Moll D, Prinz A, Gesellchen F, Drewianka S, Zimmermann B, Herberg FW: Biomolecular interaction analysis in functional proteomics. J Neural Transm 2006, 113(8):1015–1032.

8. Kraemer A, Rehmann HR, Cool RH, Theiss C, de Rooij J, Bos JL, Wittinghofer A: Dynamic interaction of cAMP with the Rap guanine-nucleotide exchange factor Epac1. J Mol Biol 2001, 306(5):1167–1177.

9. Mucignat-Caretta C, Caretta A: Binding of two fluorescent cAMP analogues to type I and II regulatory subunits of cAMP-dependent protein kinases. Biochim Biophys Acta 1997, 1357(1):81–90.

10. Sako Y, Hibino K, Miyauchi T, Miyamoto Y, Ueda M, Yanagida T: Single-molecule imaging of signaling molecules in living cells. Single Mol 2000, 1:151–155.

11. Johnson ID, Kang HC, Haugland RP: Fluorescent membrane probes incorporating dipyrrometheneboron difluoride fluorophores. Anal Biochem 1991, 198(2):228–237.

12. Haugland RP: Handbook of Fluorescent Probes and Research Products. 9th edition. Edited by: Probes GJM. Eugene, OR ; 2002.

13. Okada CY, Rechsteiner M: Introduction of macromolecules into cultured mammalian cells by osmotic lysis of pinocytic vesicles. Cell 1982, 29(1):33–41.

14. Hagen V, Bendig J, Frings S, Eckardt T, Helm S, Reuter D, Kaupp UB: Highly Efficient and Ultrafast Phototriggers for cAMP and cGMP by Using Long-Wavelength UV/Vis-Activation. Angew Chem Int Ed Engl 2001, 40(6):1045–1048.

15. Furuta T, Takeuchi H, Isozaki M, Takahashi Y, Kanehara M, Sugimoto M, Watanabe T, Noguchi K, Dore TM, Kurahashi T, Iwamura M, Tsien RY: Bhc-cNMPs as either water-soluble or membrane-permeant photoreleasable cyclic nucleotides for both one- and two-photon excitation. Chembiochem 2004, 5(8):1119–1128.

16. Tasken K, Skalhegg BS, Tasken KA, Solberg R, Knutsen HK, Levy FO, Sandberg M, Orstavik S, Larsen T, Johansen AK, Vang T, Schrader HP, Reinton NT, Torgersen KM, Hansson V, Jahnsen T: Structure, function, and regulation of human cAMP-dependent protein kinases. Adv Second Messenger Phosphoprotein Res 1997, 31:191–204.

17. Gonzalez GA, Montminy MR: Cyclic AMP stimulates somatostatin gene transcription by phosphorylation of CREB at serine 133. Cell 1989, 59(4):675–680.

18. Skalhegg BS, Tasken K: Specificity in the cAMP/PKA signaling pathway. differential expression, regulation, and subcellular localization of subunits of PKA. Front Biosci 1997, 2:d331–42.

19. Craven KB, Zagotta WN: CNG and HCN channels: two peas, one pod. Annu Rev Physiol 2006, 68:375–401.

20. de Rooij J, Rehmann H, van Triest M, Cool RH, Wittinghofer A, Bos JL: Mechanism of regulation of the Epac family of cAMP-dependent RapGEFs. J Biol Chem 2000, 275(27):20829–20836.

21. Crosby GA, Demas JN: Measurement of photoluminescence quantum yields. J Phys Chem 1971, 75(8):991–1024.

22. Melhuish WH: Quantum efficiencies of fluorescence of organic substances: effect of solvent and concentration of the fluorescent solute. J Phys Chem 1961, 65(2):229 –2235.

23. Schwede F, Christensen A, Liauw S, Hippe T, Kopperud R, Jastorff B, Doskeland SO: 8-Substituted cAMP analogues reveal marked differences in adaptability, hydrogen bonding, and charge accommodation between homologous binding sites (AI/AII and BI/BII) in cAMP kinase I and II. Biochemistry 2000, 39(30):8803–8812.

24. Ogreid D, Ekanger R, Suva RH, Miller JP, Sturm P, Corbin JD, Doskeland SO: Activation of protein kinase isozymes by cyclic nucleotide analogs used singly or in combination. Principles for optimizing the isozyme specificity of analog combinations. Eur J Biochem 1985, 150(1):219–227.

25. Ogreid D, Ekanger R, Suva RH, Miller JP, Doskeland SO: Comparison of the two classes of binding sites (A and B) of type I and type II cyclic-AMP-dependent protein kinases by using cyclic nucleotide analogs. Eur J Biochem 1989, 181(1):19–31.

26. Cook PF, Neville ME Jr., Vrana KE, Hartl FT, Roskoski R Jr.: Adenosine cyclic 3',5'-monophosphate dependent protein kinase: kinetic mechanism for the bovine skeletal muscle catalytic subunit. Biochemistry 1982, 21(23):5794–5799.

27. Dostmann WR, Taylor SS: Identifying the molecular switches that determine whether (Rp)-cAMPS functions as an antagonist or an agonist in the activation of cAMP-dependent protein kinase I. Biochemistry 1991, 30(35):8710–8716.

28. Herberg FW, Taylor SS, Dostmann WR: Active site mutations define the pathway for the cooperative activation of cAMP-dependent protein kinase. Biochemistry 1996, 35(9):2934–2942.

29. Malgaroli A, Milani D, Meldolesi J, Pozzan T: Fura-2 measurement of cytosolic free Ca2+ in monolayers and suspensions of various types of animal cells. J Cell Biol 1987, 105(5):2145–2155.

30. Poenie M, Alderton J, Steinhardt R, Tsien R: Calcium rises abruptly and briefly throughout the cell at the onset of anaphase. Science 1986, 233(4766):886–889.

31. Szaszak M, Christian F, Rosenthal W, Klussmann E: Compartmentalized cAMP signalling in regulated exocytic processes in non-neuronal cells. Cell Signal 2007.

32. Mucignat-Caretta C, Caretta A: Localization of Triton-insoluble cAMP-dependent kinase type RIbeta in rat and mouse brain. J Neurocytol 2001, 30(11):885–894.

33. Mucignat-Caretta C, Caretta A: Clustered distribution of cAMP-dependent protein kinase regulatory isoform RI alpha during the development of the rat brain. J Comp Neurol 2002, 451(4):324–333.

34. Mucignat-Caretta C, Caretta A: Distribution of insoluble cAMP-dependent kinase type RI and RII in the lizard and turtle central nervous system. Brain Res 2007.

35. Kraß J, Jastorff B, Genieser HG: Determination of Lipophilicity by Gradient Elution High-Performance Liquid Chromatography. Anal Chem 1997, 69:2575 –22581.

36. Sandberg M, Butt E, Nolte C, Fischer L, Halbrugge M, Beltman J, Jahnsen T, Genieser HG, Jastorff B, Walter U: Characterization of Sp-5,6-dichloro-1-beta-D-ribofuranosylbenzimidazole- 3',5'-monophosphorothioate (Sp-5,6-DCl-cBiMPS) as a potent and specific activator of cyclic-AMP-dependent protein kinase in cell extracts and intact cells. Biochem J 1991, 279 (Pt 2):521–527.

37. Enserink JM, Christensen AE, de Rooij J, van Triest M, Schwede F, Genieser HG, Doskeland SO, Blank JL, Bos JL: A novel Epac-specific cAMP analogue demonstrates independent regulation of Rap1 and ERK. Nat Cell Biol 2002, 4(11):901–906.

38. Ponsioen B, Zhao J, Riedl J, Zwartkruis F, van der Krogt G, Zaccolo M, Moolenaar WH, Bos JL, Jalink K: Detecting cAMP-induced Epac activation by fluorescence resonance energy transfer: Epac as a novel cAMP indicator. EMBO Rep 2004, 5(12):1176–1180.

39. Terrin A, Di Benedetto G, Pertegato V, Cheung YF, Baillie G, Lynch MJ, Elvassore N, Prinz A, Herberg FW, Houslay MD, Zaccolo M: PGE(1) stimulation of HEK293 cells generates multiple contiguous domains with different [cAMP]: role of compartmentalized phosphodiesterases. J Cell Biol 2006, 175(3):441–451.

40. Chock SP, Huang CY: An optimized continuous assay for cAMP phosphodiesterase and calmodulin. Anal Biochem 1984, 138(1):34–43.

41. Schwede F, Maronde E, Genieser H, Jastorff B: Cyclic nucleotide analogs as biochemical tools and prospective drugs. Pharmacol Ther 2000, 87(2–3):199–226.

42. Prinz A, Diskar M, Erlbruch A, Herberg FW: Novel, isotype-specific sensors for protein kinase A subunit interaction based on bioluminescence resonance energy transfer (BRET). Cell Signal 2006, 18(10):1616–1625.

43. Schultz C, Vajanaphanich M, Harootunian AT, Sammak PJ, Barrett KE, Tsien RY: Acetoxymethyl esters of phosphates, enhancement of the permeability and potency of cAMP. J Biol Chem 1993, 268(9):6316–6322.

44. Marullo S, Bouvier M: Resonance energy transfer approaches in molecular pharmacology and beyond. Trends Pharmacol Sci 2007, 28(8):362–365.

45. Slice LW, Taylor SS: Expression of the catalytic subunit of cAMP-dependent protein kinase in Escherichia coli. J Biol Chem 1989, 264(35):20940–20946.

46. Herberg FW, Bell SM, Taylor SS: Expression of the catalytic subunit of cAMP-dependent protein kinase in Escherichia coli: multiple isozymes reflect different phosphorylation states. Protein Eng 1993, 6(7):771–777.

47. Laemmli UK: Cleavage of structural proteins during the assembly of the head of bacteriophage T4. Nature 1970, 227(5259):680–685.

48. Prinz A, Diskar M, Herberg FW: Application of bioluminescence resonance energy transfer (BRET) for biomolecular interaction studies. Chembiochem 2006, 7(7):1007–1012.

Fast Benchtop Fabrication of Laminar Flow Chambers for Advanced Microscopy Techniques

David S. Courson and Ronald S. Rock

ABSTRACT

Background

Fluid handling technology is acquiring an ever more prominent place in laboratory science whether it is in simple buffer exchange systems, perfusion chambers, or advanced microfluidic devices. Many of these applications remain the providence of laboratories at large institutions with a great deal of expertise and specialized equipment. Even with the expansion of these techniques, limitations remain that frequently prevent the coupling of controlled fluid flow with other technologies, such as coupling microfluidics and high-resolution position and force measurements by optical trapping microscopy.

Method

Here we present a method for fabrication of multiple-input laminar flow devices that are optically clear [glass] on each face, chemically inert, reusable, inexpensive, and can be fabricated on the benchtop in approximately one hour. Further these devices are designed to allow flow regulation by a simple gravity method thus requiring no specialized equipment to drive flow. Here we use these devices to perform total internal reflection fluorescence microscopy measurements as well as position sensitive optical trapping experiments.

Significance

Flow chamber technology needs to be more accessible to the general scientific community. The method presented here is versatile and robust. These devices use standard slides and coverslips making them compatible with nearly all types and models of light microscopes. These devices meet the needs of groups doing advanced optical trapping experiments, but could also be adapted by nearly any lab that has a function for solution flow coupled with microscopy.

Introduction

Coupling of optical trapping with flow chambers, lab-on-a-chip, and other microfluidic devices has been accomplished by a number of groups[1]–[6] for two major purposes; cell sorting and manipulation[1]–[4] and making biochemical and biophysical measurements[5], [6]. Most of the cell sorting devices use traditional fabrication methods and have one optically uniform glass face while the channel walls and opposing face are made of a polymer such as polydimethylsiloxane (PDMS)[7].

Traditional microfluidic fabrication methods have several limitations. First, the fabrication method requires specialized equipment not readily available to many groups. Second, the PDMS face opposite the coverslip prevents some optical techniques that require accurate visualization through both sides of the device. One such technique is optical trapping with precision force and position measurements, because the PDMS layers are of non-uniform density and distort the wavefront of the required detection lasers in unpredictable ways. This can make experiments such as the study of molecular motor stepping[8], [9] or forced unfolding[10] problematic in such devices. Since buffer conditions can change these behaviors, being able to quickly change the buffer conditions while studying the same molecules could be useful but remains illusive.

As a result of these difficulties most trapping experiments that require precision measurements are performed using chambers made of a slide and coverslip linked by a piece of double sided tape and sealed with vacuum grease[11]. When multiple conditions are required, multiple independent experiments are run. While very effective these devices have limitations as well. Experiments using sealed sticky tape devices require premixing all components, so unwanted component self-association prior to visualization is a problem. It is also difficult to make complex geometry devices using tape flow cells. Further, coupling tape devices to flow often requires drilling through glass and attaching ports, which can interfere with optical components such as oil condensers[12].

For biochemical and biophysical measurements with flow other devices have been produced. The Kowalczykowski group has coupled optical trapping and flow technology to generate a platform for performing fluorescence microscopy measurements on proteins bound to a single DNA molecule[5]. However these devices are made from etched glass, which requires special facilities to fabricate and have not been, to our knowledge, coupled to high-resolution detection systems.

Recently the Wuite group developed a system to couple flow and optical trapping with high-resolution position and force detection[6] to examine the single molecule behavior of DNA binding proteins. They report using parafilm to construct channels. In our hands uniform parafilm adherence was problematic so these devices tended to leak. We also had difficulties incorporating fluid ports that would not interfere with our optics. Our research requires a device of this type that prevents reagent mixing and self-assembly but allows trapped particles to be transported into different environments. Here we present a simple manual method for flow chamber fabrication that can be preformed on a benchtop and is fast, inexpensive, requires no glass drilling, produces devices that are reusable and meet the requirements of advanced optical trapping assays, including all-glass optical faces, and a radial arrangement of tubing connections that allows us to use oil-immersion condensers and objectives.

Methods

Device Fabrication

Our flow chamber design uses a single flat sheet of silicon rubber, which is cut by hand into the desired shape. The advantage of using commercially available silicon rubber sheets is that they have a uniform thickness. Moreover, they are easily cut by hand and manipulated, since they only adhere once activated by oxygen plasma. We cut out a piece of silicon rubber (McMaster-Carr, 87315K63) of the same dimensions as the coverslips that will be used (25 mm round). Next the desired pattern is cut into a silicon rubber sheet using sharp razors (Pattern in Figure 1a). Channel entrances should be left intact at this point so that sheet remains in one piece until after it is joined to the slide (Figure 1). The cut sheet constitutes the device body.

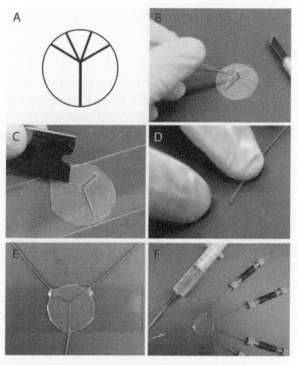

Figure 1. Fabrication and use of devices. A) Scale pattern of a four input device made to fit a 25 mm round coverslip. B) Device body of a two input device being prepared. The channel pattern is cut out of the body but the channel entrances are left intact so the device remains one solid piece during the process of cleaning and adhering to the slide. C) Channel entrances are cut open with a razor blade. A scraping motion is needed to remove the piece once cuts are made. D) Pieces of PEEK tubing are cut and one end is flattened to fit into the device. E) Tubing is attached using epoxy. The device is complete at this point. F) Example of a functioning four input device attached via HPLC connectors to a reservoir and to a priming syringe via silicone tubing. These linkages can be modified to fit other systems.

The microscope slide and the device body should be rinsed with pure, deionized water and dried to remove large debris. Plasma clean the slide and device body to functionalize the surfaces (acid washing will also work eliminating the need for a plasma cleaner[13]). Once cleaned press the device body onto the slide. The bond is permanent. Cut channel entrances open with the razor.

Next plasma clean the device complex and coverslip and press them together. This produces a sandwich of rubber with channel openings between the slide and coverslip. Thin wall PEEK tubing (Upchurch, 1569) is then connected to the device. Compress tubing at one end and slide into channel entrances. Seal the joint between the device and the tubing with quick drying epoxy. Once dry the device is complete and can be hooked up to a flow driving system via HPLC adaptors. All devices should be tested for leaks and correct flow characteristics before use in experiments.

Device Testing

Since the devices are not precision cut or cast, there will be variance from device to device. Thus, if very specific or complex flow characteristics are required this technique may not be suitable. Additionally, each flow cell should be characterized individually. For example, in multiple input laminar flow cells, if there is variance in the structure of the input channels or the length of the tubing leading from the reservoir the width and flow rate of each lane in the main channel may vary slightly. Even with these limitations for many applications this method is sufficient.

Flow rates can be measured in several ways. For high flow rates [>1 ul/sec] driving a set amount of fluid through the device and measuring the time required to pass a given volume is an effect method. For extremely slow flow rates, observing the motion of particles such as fluorescent beads as they transit a known distance as visualized by the microscope can give an accurate estimation of flow rate. It is important to note that flow rates are fastest in the center of the channels and slower near surfaces. If driving flow by gravity it is important to remember that rates will drop as the column heights balance. Once at a desired flow rate, the water column height must be routinely adjusted to maintain the flow rate.

Device Cleaning

Since these systems are entirely composed of PEEK, silicon rubber and glass they are very chemically inert. As such they can be cleaned very rigorously allowing them to be reused without fear of contamination. We rinse the systems sequentially with water, 1 M NaOH, EtOH then finally water again. This cleans all internal surfaces.

Myosin Motility Assay

This assay was performed in two ways: in a flow chamber and a series of sealed chambers (the current standard method). Data in panels 3b and 3c were collected on different days but used the same motor preparation and solution stocks. Incubation times and all other variables were kept as similar as possible.

For the flow chamber experiments, a four input laminar flow chamber was used. The device was primed with water by injecting it into the device via the output port using a syringe. All setup solutions are added in this way. This eliminates bubbles and insures all inputs are exposed to the same conditions. Anti-GFP antibody (QBiogene, 50 ng/µl) was added into the device to coat all of the internal surfaces[14]. BSA was then added as a blocking agent (1 mg/ml). GFP tagged Myosin VI heavy meromyosin (HMM) dimer [14] was added followed by filamentous actin (chicken skeletal muscle, 100 nM [15]), and then assay buffer (25 mM imidazole, pH 7.5, 25 mM KCl, 1 mM EGTA, 4 mM MgCl2, 10 mM DTT, 0.86 mg/ml glucose oxidase, 0.14 mg/ml catalase, 9 mg/ml glucose) with zero ATP. The microscope was focused and a desirable field of view was found. Assay buffers with desired ATP concentrations (0, 0.1, 1, 2 µM) were loaded into the input reservoirs. Movies were then recorded for the zero ATP condition. The valve to the 0.1 µM ATP containing solution was then opened, a sufficient replacement volume was allowed to flow through and a movie was recorded. Data in panel 3b were recorded with the flow turned off. Data in panel 3c were recorded with flow off then with flow on. This process was repeated for each ATP solution. Data was recorded on a home built total internal reflection fluorescence (TIRF) microscope with Andor iXon camera. Movies were analyzed using ImageJ.

Optical Trapping Assay

All experiments were performed on a home built fluorescence and optical trapping microscope.

Power Spectra

A two input device was used. The system was primed as described above. 1 µm beads were loaded into one input reservoir, and assay buffer into the other. Beads were flowed into the device and a single bead was trapped. Flow rate was approximately 1 µL/sec. The trapped bead was moved into the buffer channel and the bead position was recorded at 10 kHz, without a lowpass antialias filter. Flow was then turned off and the power spectrum was repeated with the same bead. Power spectra were calculated using the Igor Pro software package.

Dumbbell

A four input device was used. BSA blocking solution was pumped into the output port of the device as described above to prime the device and block the surface. 1 µm neutravidin coated beads (biotinylated beads from Invitrogen incubated in a neutravidin solution and blocked with BSA) were loaded into one input reservoir,

TMR-phalloidin (Sigma-Aldrich) stabilized actin filaments with ten percent of the monomers labeled with biotin were loaded into the second input reservoir, and buffer was loaded into the third and fourth. The solutions were allowed to flow into the device. Beads were trapped then the stage was moved so that those beads transited into the actin containing channel. Once an actin filament was captured (by flowing against a bead) the stage was again moved so the beads and captured filament moved into the buffer channel where the dumbbell was assembled. The bead with the attached filament was placed upstream. Buffer flow extended the filament. The second bead was then moved into position at the free end of the extended filament and the bond was formed.

Results

Here we present data from several experiments that show the broad applicability and high degree of functionality of these devices. This data shows that these devices are functional for a broad range of microscopy experiments including sensitive optical trapping experiments.

Device Characteristics

Devices were fabricated using main channel widths of 0.3 up to 2 mm, input channel widths of 0.5 to 1.0 mm, and between 1 and 5 inputs. All devices produced laminar flow, where each solution in the main channel remained separate and only mixed via diffusion (Figure 2). Devices were tested over a range of flow rates from greater than a 1 milliliter per second to rate of below 1 microliter per minute.

Myosin Motility Assay with Multiple ATP Concentrations

As a proof-of-principle experiment these devices were used to perform myosin gliding filament assays under varying ATP concentrations using TIRF microscopy. Myosin VI (as is true of all myosins) shows an ATP concentration dependent motility rate [16], [17]. The device was coated with myosin VI before actin filaments were added. Some filaments were bound by the motors at the coverslip surface and immobilized in the absence of ATP. A series of motility buffers with increasing ATP concentrations (0, 0.1, 1, 2 µM) was then allowed to flow into the flow chamber and movies were recorded at each ATP concentration, with and without flow. The addition of ATP caused the motors to move (Figure 3). The motility profiles were similar between the two assays.

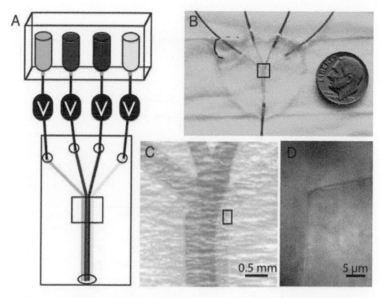

Figure 2. Device Details and Flow Characteristics. A) Diagram of basic flow chamber design. Solutions are stored in reservoirs and must pass through a solenoid valve to reach the flow chamber. Solutions can be added one by one or multiple at once. When multiple solutions are added they flow in separate laminar flow lanes. This behavior is preserved regardless of the number of solutions entering the flow chamber. All the solution flows out through a single output port. B) Picture of a four input flow chamber with four solutions flowing. Each input has a different colored water solution flowing through it to demonstrate laminar flow characteristics of the device. The device is made with a standard microscope slide and 25 mm round #1.5 coverslip. Orange PEEK tubing is connected to each input and the output with epoxy. The output tube leads to a collection reservoir mounted on an adjustable height platform to allow control of the flow rate by altering the height of the water column difference between the input reservoirs and the output tubing end. C) Blow up of flow chamber from highlighted area in panel B, showing laminar flow characteristics. Small defects such as the one on the leftmost corner do not significantly affect the downstream flow profile. D) Blow up from inside the highlighted area in panel C. Bright field image of a feature in the device. This image shows the corner where one of the input channels joins the main channel. The scale bar is set over the rubber section while the figure letter is set over the input channel solution. Even though these devices are cut by hand they typically have very fine and clean features so solution flow is not perturbed.

There are several advantages to the flow chamber method. This method allowed us to assess the ATP-dependent motility of a single population of motors, rather than preparing a different slide to analyze each ATP concentration. This prevents slide-to-slide variation, making results easy and reliable to interpret. Open chamber devices (sticky tape) can dry out during the course of an experiment as happens if inadequately sealed, changing ATP concentration. The flow chambers presented here do not share this limitation. Finally, continuous solution flow can also be used to prevent the accumulation of ADP in the chamber, which could lower the apparent motility rate of the motor. The shear induced by flow does not appear to affect myosin VI motility but it should be considered if assaying other motors in the presence of flow.

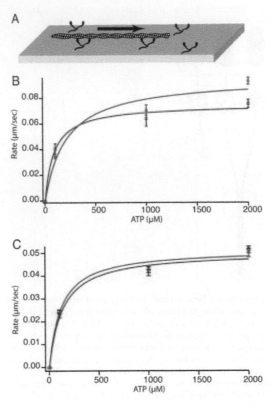

Figure 3. Myosin VI Motility. A) Motility assay cartoon. Myosin VI motors are immobilized on a coverslip surface. Actin filaments land on the surface and are then propelled by the motor proteins. B) Graph of motility rate verses concentration of ATP. Data in red (squares) indicates experiments performed in separate sealed chambers. Data in blue (circles) indicates a single experiment performed in a flowcell device where a desired ATP solution was flowed into the chamber, flow was stopped and data was recorded for each concentration. Michaelis-Menton fit for each data set in corresponding colors. Fits are similar though some difference is seen at 100 μM and 2 mM. Fit parameters for sealed chambers: Vmax = .075±.0040 μm/sec, Km = 88±25 μM. Fit parameters for flowcells: Vmax = .096±.0096 μm/sec, Km = 206±95 μM. C) Graph of motility rate verses concentration of ATP. Comparing rates obtained in a single flow chamber with flow on (red cross) and off (blue circles). Flow rate used was approximately 1 μl/sec. The fits are nearly identical, indicating surface shear experienced by the motor when flow is on had no measurable affect of motor behavior. Fit parameters with flow: Vmax = .051±.0030 μm/sec, Km = 129±38 μM. Fit parameters without flow: Vmax = .051±.0031 μm/sec, Km = 114±35 μM. Indeed, variation from day-to-day or sample-to-sample is far higher than variation caused by flow shear, as is evident by the difference in behavior between samples in panel B and panel C.

Optical Trapping Assays

To demonstrate that optical trapping, manipulation, and position detection is practical with these devices two assays were performed. 1) We trapped 1 μm polystyrene beads and measured power spectrums with and without flow. 2) We assembled actin dumbbells and manipulated them. The first experiment uses optical trapping, bright field microscopy and laser position detection. The second uses optical trapping, bright field and epifluorescent microscopy.

Power Spectra

A bead was trapped under flow of 1 μL/sec flow rate. A power spectrum was recorded [8], [18], [19]. Then flow was stopped and another power spectrum was measured. The power spectra are nearly indistinguishable in the X direction (the direction of flow) (Figure 4). First this demonstrates that trapping is indeed possible in these devices. Second these devices are functional for position detection and force measurements when the system is under flow and when flow is stopped. This creates a dynamic environment in which to design and perform experiments.

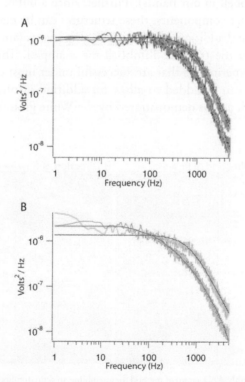

Figure 4. Power Spectra. A) Representative power spectrum for a trapped bead in a flow chamber with flow turned off. A spectrum is taken in X and Y coordinate systems. X is shown in red, Y in purple. Lorentzian fits are overlaid on the each spectrum. The black fit is for X, the blue fit is for Y. The bead corner frequency for X = 781 Hz, for Y = 472 Hz. B) Power spectrum data for the same bead after the flow was turned on at a rate of 1 μl/sec. Flow is in the X direction. X is shown in orange, Y in green. The black fit is for X, the purple fit is for Y. The bead corner frequency for X = 697 Hz, for Y = 226 Hz. Introduction of flow causes some low frequency noise in both spectra and shifts the corner frequency of the Y power spectra with respect to the no flow. X spectra are very similar for each condition.

Dumbbells

For many actin based optical trapping experiments, such as analysis of myosin motors stepping, forming an actin dumbbell is required [20], [21]. A dumbbell is an actin filament with each end connected to beads held in separate optical traps. Here we trapped two beads under flow, attached an actin filament to one of those beads under flow, then moved the second bead into position downstream behind the trapped filament to form the dumbbell (Figure 5). The trapped filament can be manipulated and used for desired assays.

Unlike traditional methods where beads and actin filaments are mixed together in extremely low dilutions and the user must find and assemble them (a process which can be time sensitive), using the flow based method present here each component is added in isolation and can be assembled extremely quickly (as fast as 13 seconds for a dumbbell in our hands). Further, since a buffer channel is present which lacks any other components, these structures can be manipulated without concern for accidental addition of unwanted components (another bead or actin filament sticking to the trapped dumbbell for example). This ability increases the percentage of experiments that are successful under most circumstances. Additional flow lanes can be added to allow for addition of other components to the trapped scaffold, as was demonstrated by the Wuite group's DNA and H-NS experiments[6].

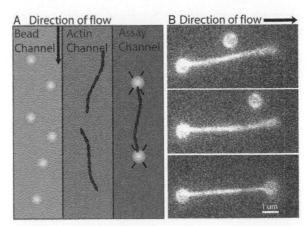

Figure 5. Dumbbell Assembly. A) Diagram of method for assembling an actin dumbbell while under flow. Beads are trapped in the bead channel, the stage is moved transferring the trapped beads into the actin channel where one filament is attached to one of the beads. The stage is moved again transferring the trapped beads and actin filament into the assay chamber where a dumbbell is formed and manipulated. B) A trapped 1 μm neutravidin coated bead with and attached biotinylated filament is held in an optical trap. Flow keeps the filament aligned down stream. A second trapped bead is then moved into position near the free end of the filament. When the bead and filament touch they become tightly linked. Introducing the components in separate channels has the advantage that the beads and filaments are isolated and do not spontaneously assemble into undesired aggregates.

Discussion

Design and Potential Applications

Beyond the ability to perform specialized optical trapping assays these devices have other important virtues. The fabrication technique reported here requires no glass drilling, no training, and little to no specialized equipment (plasma cleaner is recommended). Devices are constructed using standard microscope slides and coverslips making them compatible with nearly all light microscopes. The assembly protocol is general enough that any glass coverslip can be accommodated, allowing for matching of coverslips to objectives for optimum results and making design of devices of different dimensions simple. The slide can even be replaced by a second coverslip to accommodate dual objective optical trapping microscopes. The fluid volume of the devices is also flexible ranging from 30 down to 3 microliters depending on design. After most usages devices can be cleaned and reused. Because of the uniform nature of the silicone rubber sheets used to make these devices, when the devices are properly assembled the coverslips lay extremely flat allowing for minimal focal drift as the sample is moved. All fluorescence imaging characteristics seem to closely mimic those of sealed sticky tape flow cells, including low oxygen permeability yielding long fluorescence lifetimes and the ability to clearly resolve single molecules.

These devices do carry inherent limitations, which should not be overlooked. Since channels are cut by hand variation from device to device will occur. For some applications this will require that each device be tested before use. Complex multilayer applications such as those requiring pneumatic microvalves [22], [23] cannot be adapted to work with these devices. Finally device volumes are typically around 10 microliters and require a reservoir to generate flow, so applications that use very small volumes are not practical.

We see several areas in which these devices could be immediately applicable. The flexibility in device design, low cost, and the ease of construction should allow them to be adapted by groups in many fields. Laminar flow chambers, from single to many inputs, can be made exactly as described here. Coupling valves to the fluid handling system allows for many possibilities from side-by-side laminar flow to iterative additions of different solutions.

Coupled with valves these devices allow for rapid and repeated changes in buffer conditions. With computer-controlled valves, experiments or condition screens could be automated saving researchers time and effort. Our system uses commercially available solenoid valves (Lee Company) to control inputs.

Most current optical trapping experiments use sealed chambers. This necessitates mixing all experimental components together before starting the experiment.

If these components interact with each other in solution then it becomes difficult to control and study these interactions. Devices like these allow for iterative assembly of complex systems, opening up new avenues of exploration for single molecule scientists[6]. As our knowledge increases study of more complex systems becomes necessary. These devices should serve as a useful tool for scientist attempting to do that.

Acknowledgements

The authors would like to thank Crista Brawley for preparing the myosin VI used in the motility assays. DSC would like to thank Wesley Jun and Stas Nagy for the use of their camera equipment.

Authors' Contributions

Conceived and designed the experiments: DSC RSR. Performed the experiments: DSC. Analyzed the data: DSC RSR. Contributed reagents/materials/analysis tools: DSC RSR. Wrote the paper: DSC.

References

1. Enger J, Goksor M, Ramser K, Hagberg P, Hanstorp D (2004) Optical tweezers applied to a microfluidic system. Lab Chip 4: 196–200.

2. Perroud TD, Kaiser JN, Sy JC, Lane TW, Branda CS, et al. (2008) Microfluidic-based cell sorting of Francisella tularensis infected macrophages using optical forces. Anal Chem 80: 6365–72.

3. Murata M, Okamoto Y, Park YS, Kaji N, Tokeshi M, et al. (2009) Cell separation by the combination of microfluidics and optical trapping force on a microchip. Anal Bioanal Chem.

4. Roman GT, Chen Y, Viberg P, Culbertson AH, Culbertson CT (2007) Single-cell manipulation and analysis using microfluidic devices. Anal Bioanal Chem 387: 9–12.

5. Galletto R, Amitani I, Baskin RJ, Kowalczykowski SC (2006) Direct observation of individual RecA filaments assembling on single DNA molecules. Nature 443: 875–8.

6. Dame RT, Noom MC, Wuite GJ (2006) Bacterial chromatin organization by H-NS protein unravelled using dual DNA manipulation. Nature 444: 387–90.

7. McDonald JC, Duffy DC, Anderson JR, Chiu DT, Wu H, et al. (2000) Fabrication of microfluidic systems in poly(dimethylsiloxane). Electrophoresis 21: 27–40.

8. Neuman KC, Block SM (2004) Optical trapping. Rev Sci Instrum 75: 2787–809.

9. Visscher K, Schnitzer MJ, Block SM (1999) Single kinesin molecules studied with a molecular force clamp. Nature 400: 184–9.

10. Kellermayer MS, Smith SB, Granzier HL, Bustamante C (1997) Folding-unfolding transitions in single titin molecules characterized with laser tweezers. Science 276: 1112–6.

11. Kron SJ, Spudich JA (1986) Fluorescent actin filaments move on myosin fixed to a glass surface. Proc Natl Acad Sci USA 83: 6272–6.

12. Brewer LR, Bianco PR (2008) Laminar flow cells for single-molecule studies of DNA-protein interactions. Nature Methods 5: 517–25.

13. Campbell K, Groisman A (2007) Generation of complex concentration profiles in microchannels in a logarithmically small number of steps. Lab Chip 7: 264–72.

14. Rock RS, Ramamurthy B, Dunn AR, Beccafico S, Rami BR, et al. (2005) A flexible domain is essential for the large step size and processivity of myosin VI. Mol Cell 17: 603–9.

15. Pardee JD, Spudich JA (1982) Purification of muscle actin. Meth Enzymol 85: 164–81.

16. Rock RS, Rice SE, Wells AL, Purcell TJ, Spudich JA, et al. (2001) Myosin VI is a processive motor with a large step size. Proc Natl Acad Sci USA 98: 13655–9.

17. Sellers JR (2000) Myosins: a diverse superfamily. Biochim Biophys Acta 1496: 3–22.

18. Svoboda K, Schmidt CF, Schnapp BJ, Block SM (1993) Direct observation of kinesin stepping by optical trapping interferometry. Nature 365: 721–7.

19. Mehta AD, Finer JT, Spudich JA (1997) Detection of single-molecule interactions using correlated thermal diffusion. Proc Natl Acad Sci USA 94: 7927–31.

20. Finer JT, Simmons RM, Spudich JA (1994) Single myosin molecule mechanics: piconewton forces and nanometre steps. Nature 368: 113–9.

21. Rock RS, Rief M, Mehta AD, Spudich JA (2000) In vitro assays of processive myosin motors. Methods 22: 373–81.

22. Studer V, Hang G, Pandolfi A, Ortiz M, Anderson WF, et al. (2004) Scaling properties of a low-actuation pressure microfluidic valve. Journal of Applied Physics 95: 393–398.

23. Oh KW, Ahn CH (2006) A review of microvalves. J Micromech Microeng 16: R13–R39.

A New Classification System for Bacterial Rieske Non-Heme Iron Aromatic Ring-Hydroxylating Oxygenases

Ohgew Kweon, Seong-Jae Kim, Songjoon Baek,
Jong-Chan Chae, Michael D. Adjei, Dong-Heon Baek,
Young-Chang Kim and Carl E. Cerniglia

ABSTRACT

Background

Rieske non-heme iron aromatic ring-hydroxylating oxygenases (RHOs) are multi-component enzyme systems that are remarkably diverse in bacteria isolated from diverse habitats. Since the first classification in 1990, there has been a need to devise a new classification scheme for these enzymes because many RHOs have been discovered, which do not belong to any group in the previous classification. Here, we present a scheme for classification of RHOs reflecting new sequence information and interactions between RHO enzyme components.

Result

We have analyzed a total of 130 RHO enzymes in which 25 well-characterized RHO enzymes were used as standards to test our hypothesis for the proposed classification system. From the sequence analysis of electron transport chain (ETC) components of the standard RHOs, we extracted classification keys that reflect not only the phylogenetic affiliation within each component but also relationship among components. Oxygenase components of standard RHOs were phylogenetically classified into 10 groups with the classification keys derived from ETC components. This phylogenetic classification scheme was converted to a new systematic classification consisting of 5 distinct types. The new classification system was statistically examined to justify its stability. Type I represents two-component RHO systems that consist of an oxygenase and an FNRC-type reductase. Type II contains other two-component RHO systems that consist of an oxygenase and an FNRN-type reductase. Type III represents a group of three-component RHO systems that consist of an oxygenase, a [2Fe-2S]-type ferredoxin and an FNRN-type reductase. Type IV represents another three-component systems that consist of oxygenase, [2Fe-2S]-type ferredoxin and GR-type reductase. Type V represents another different three-component systems that consist of an oxygenase, a [3Fe-4S]-type ferredoxin and a GR-type reductase.

Conclusion

The new classification system provides the following features. First, the new classification system analyzes RHO enzymes as a whole. RwithSecond, the new classification system is not static but responds dynamically to the growing pool of RHO enzymes. Third, our classification can be applied reliably to the classification of incomplete RHOs. Fourth, the classification has direct applicability to experimental work. Fifth, the system provides new insights into the evolution of RHO systems based on enzyme interaction.

Background

Microorganisms play indispensable roles in the degradation and detoxification of polycyclic aromatic hydrocarbons (PAHs) in the environment [1,2]. The initiation of the aerobic microbial degradation of PAHs is an oxidative attack [3,4]. The enzymes that catalyze insertion of molecular oxygen into aromatic benzene rings are termed oxygenases [5]. They require transition metals, such as iron and heme, as catalytic centers. Oxygenases that utilize non-heme Fe(II) are called

Rieske-type non-heme iron aromatic ring-hydroxylating oxygenase (RHO) whereas others that use heme are cytochrome P450s [6,7]. The term RHO is used herein to denote the Rieske-type non-heme iron ring-hydroxylating oxygenase.

Although RHOs mostly use NAD(P)H as an electron donor and catalyze the same oxygenation reaction, they are remarkably diverse with respect to their structure [3,4,8]. RHOs are multi-component enzymes of two or three protein components consisting of an electron transport chain (ETC) and an oxygenase. Oxygenase components are either homo- (αn) or hetero-oligomers (αnβn) and in each case, the α subunit, called large subunit, contains two conserved regions, a Rieske [2Fe-2S] center and non-heme mononuclear iron. The α subunits are known to be the catalytic components involved in the transfer of electrons to oxygen molecules. The ETC that transfers reducing equivalents from NAD(P)H to the oxygenase components consists of either a flavoprotein reductase or a flavoprotein reductase and a ferredoxin [3,4]. An interaction between oxygenase and ETC components is required for the enzyme system to transfer electrons from the electron donor to aromatic hydrocarbon electron acceptor. The RHO enzyme system has been extensively studied in many different microorganisms since the initial reaction mostly determines the aromatic substrate for degradation [9-15].

Classification of RHOs is essentially an effort to organize the information into a system that is useful for understanding the relationship between various aspects of sequence, structure, function and evolution. A three-class system (class I, II and III) was initially instituted by Batie et al. [16]. Based on the number of constituent components and the nature of the redox centers, this classification was able to give systematic information about RHOs. We will refer to this approach as "the traditional classification". With the recent tremendous accumulation of new sequence information on RHOs, there is a current need for a new classification strategy that can transform the multitude of complex data into useful organized information. In this regard, computational phylogenetic analysis of molecular sequence was imperative, which we term "the phylogenetic classification". Several challenges have been introduced using this method. Werlen et al. [17] grouped RHOs into four families based on substrate specificities and sequence alignments with associated distance calculations. This classification emphasized the structure-function relationship of the oxygenase component. However, some RHOs appear not to fit in this scheme probably because of the small RHO sample pool which resulted in the partial observation. Nam et al. [18] also proposed a clustering system based on the homology of the amino acid sequences of terminal oxygenase components. This classification system was more inclusive and well reflected the phylogenetic affiliation among RHOs.

In recent years, we characterized 3 oxygenases, NidAB, NidA3B3 and PhtAaAb, from M. vanbaalenii PYR-1 involved in the oxidation of aromatic hydrocarbons

[9,10,15,19]. Phylogenetic analysis of these oxygenase components, NidA, NidA3 and PhtAa, showed that they clustered with a new group of α subunits found in Nocardioides, Rhodococcus, Terrabacter, Arthrobacter and other Mycobacterium spp. [20-25]. Interestingly, RHO genes found in this group of bacteria appear to have features in common; genes for ETC components are not always closely positioned with oxygenase genes (genetic discreteness) and limited numbers of ferredoxin and reductase components are shared by multiple oxygenases (numerical imbalance). Interestingly, ferredoxin components, that are compatible with the oxygenase enzymes from this bacterial group, were often identified to be a [3Fe-4S] cluster type [20,22,23]. The [3Fe-4S] type of ferredoxin has been recently introduced as an ETC component for RHO enzyme systems, the other ferredoxin components being the [2Fe-2S] type. In fact, NidAB and NidA3B3 from PYR-1 were also shown to be compatible with the [3Fe-4S]-type ferredoxin, PhtAc and ferredoxin reductase (PhtAd) [19]. However, the classification system proposed by Batie et al. and Nam et al. can not explain these new RHO features, the [3Fe-4S]-type ferredoxin, genetic discreteness of ETC components and the numerical imbalance between oxygenase and ETC component. In addition, considering that RHOs are multi-component enzyme systems, ETC components along with oxygenases components are also necessary for their functional understanding. Therefore, the new classification system was launched with the aim of analyzing the RHO components as a whole. The classification system not only systematically incorporates each component data but also basically reflects the previous "traditional" and "phylogenetic" classification.

Methods

Sequences Retrieving

The amino acid sequences of RHO enzymes were retrieved by BLAST searches from public databases including NCBI nonredundant (NR) protein database [26]. From several hundreds of RHO enzymes that were recovered, we first ended up with the selection of 130 RHO enzymes. Among them, 25 "standard" RHO enzymes were chosen (Table 1) and used to construct the new classification system. These standard RHO enzymes are all well known with respect to genetics and enzyme functions. We also took sequence identities among RHO enzymes into consideration for the selection of standard RHO samples, in which highly identical RHOs were only once selected, so as for the samples to be a full representative of RHO enzymes. The size of the 25 standard RHOs as well as their quality as an RHO representative were statistically examined. To evaluate the new classification scheme, 38 RHO samples (Table 2) were selected as test enzyme systems. For some of these test enzymes, the genetic information for both the oxygenase and

ETC components are available whereas for others only oxygenase information is available. To apply the new classification system, a total of 130 RHO enzymes, which include additional 67 RHO enzymes, were classified.

Table 1. Standard RHO enzymes used in this study.

Standard RHO enzyme system	Gene	Oxygenase Prosthetic group	Ferredoxin Prosthetic group	Reductase Prosthetic group	Structure	Accession number
Carbazole 1,9a-dioxygenase (*P. resinovorans* CA10) [44]	carAa	[2Fe-2S]/Fe^{2+}	[2Fe-2S]	FAD/[2Fe-2S]	FNR$_N$-type	D89064
Phenoxybenzoate dioxygenase (*Alcaligenes sp.* BR60) [45]	cbaA	[2Fe-2S]/Fe^{2+}	None	FMN/[2Fe-2S]	FNR$_C$-type	U18133
Phenoxybenzoate dioxygenase (*P. pseudoalcaligenes* POB310) [46]	pobA	[2Fe-2S]/Fe^{2+}	None	FMN/[2Fe-2S]	FNR$_C$-type	X78823
Phthalate dioxygenase (*B. cepacoa* DBO1) [13]	ophA2	[2Fe-2S]/Fe^{2+}	None	FMN/[2Fe-2S]	FNR$_C$-type	AF095748
p-Toluenesulfonate monooxygenase (*C. testosterone* T-2) [47]	tsaM	[2Fe-2S]/Fe^{2+}	None	FMN/[2Fe-2S]	FNR$_C$-type	U32622
Aniline dioxygenase (*Acinetobacter sp.* YAA) [48]	atdA	[2Fe-2S]/Fe^{2+}	None	FAD/[2Fe-2S]	FNR$_C$-type	D86080
Aniline oxygenase (*P. putida* UCC22) [49]	tdnA1	[2Fe-2S]/Fe^{2+}	None	FAD/[2Fe-2S]	FNR$_C$-type	D85415
2-Halobenzoate 1,2-dioxygenase (*P. cepacia* 2CBS) [50]	cbdA	[2Fe-2S]/Fe^{2+}	None	FAD/[2Fe-2S]	FNR$_N$-type	X79076
Benzoate 1,2-dioxygenase (*Acinetobacter.* sp. ADP1) [51]	BenA	[2Fe-2S]/Fe^{2+}	None	FAD/[2Fe-2S]	FNR$_N$-type	AF009224
Anthranilate dioxygenase (*Acinetobacter.* sp. ADP1) [51]	antA	[2Fe-2S]/Fe^{2+}	None	FAD/[2Fe-2S]	FNR$_N$-type	AF071556
Phenanthrene dioxygenase (*Nocardioides sp.* KP7) [11]	phdA	[2Fe-2S]/Fe^{2+}	[3Fe-4S]	FAD	GR-type	AB017795
Phthalate dioxygenase (*Terrabacter sp.* DBF63) [22]	phtA1	[2Fe-2S]/Fe^{2+}	[3Fe-4S]	FAD	GR-type	AB084235
Phthalate dioxygenase (*A. keyseri* 12B) [23]	phtAa	[2Fe-2S]/Fe^{2+}	[3Fe-4S]	FAD	GR-type	AF331043
Phthalate dioxygenase (*M. vanbaalenii* PYR-1) [10]	phtAa	[2Fe-2S]/Fe^{2+}	[3Fe-4S]	FAD	GR-type	AY365117
Phthalate dioxygenase (*Rhodococcus sp.* RHA1) [34]	padAa2	[2Fe-2S]/Fe^{2+}	[3Fe-4S]	FAD	GR-type	AB154537
3,4-Dihydroxyphenanthrene dioxygenase (*A. foecalis* AFK2) [52]	phnAc	[2Fe-2S]/Fe^{2+}	[2Fe-2S]	FAD/[2Fe-2S]	FNR$_N$-type	AB024945
Naphthalene dioxygenase (*Pseudomonas sp.* 9816-4) [12]	nahAc	[2Fe-2S]/Fe^{2+}	[2Fe-2S]	FAD/[2Fe-2S]	FNR$_N$-type	U49496
PAH dioxygenase (*P. putida* OUS82) [53]	puhAc	[2Fe-2S]/Fe^{2+}	[2Fe-2S]	FAD/[2Fe-2S]	FNR$_N$-type	AB004059
Carbazole dioxygenase (*Sphingomonas sp.* CB3) [54]	carAa	[2Fe-2S]/Fe^{2+}	[2Fe-2S]	FAD	GR-type	AF060489
Dioxin dioxygenase (*Sphingomonas sp.* RW1) [14]	dxnA1	[2Fe-2S]/Fe^{2+}	[2Fe-2S]	FAD	GR-type	X72850
Biphenyl dioxygenase (*Rhodococcus sp.* RHA1) [55]	bphA1	[2Fe-2S]/Fe^{2+}	[2Fe-2S]	FAD	GR-type	D32142
Toluene dioxygenase (*P. putida* F1) [56]	todC1	[2Fe-2S]/Fe^{2+}	[2Fe-2S]	FAD	GR-type	J04996
Biphenyl dioxygenase (*Pseudomonas sp.* KKS102) [57]	bphA1	[2Fe-2S]/Fe^{2+}	[2Fe-2S]	FAD	GR-type	D17319
Biphenyl 2,3-dioxygenase (*B. xenovorans* LB400) [58]	bphA	[2Fe-2S]/Fe^{2+}	[2Fe-2S]	FAD	GR-type	M86348
Biphenyl dioxygenase (*P. pseudoalcaligenes* KF707) [59]	bphA1	[2Fe-2S]/Fe^{2+}	[2Fe-2S]	FAD	GR-type	M83673

Table 2. Test RHO enzymes used in this study.

Test RHO enzyme system	Gene	Oxygenase Prosthetic group	Ferredoxin Prosthetic group	Reductase Prosthetic group	Structure	Accession number
PAH dioxygenase (*M. vanbaalenii* PYR-1) [9]	orf25	[2Fe-2S]/Fe^2	unknown	unknown	unknown	AY365117
	nidA	[2Fe-2S]/Fe^2	unknown	unknown	unknown	AF249301
	nidA3	[2Fe-2S]/Fe^2	unknown	unknown	unknown	DQ028634
PAH dioxygenase (*Mycobacterium sp.* 6PY1) [25]	pdoA1	[2Fe-2S]/Fe^2	unknown	unknown	unknown	AJ494745
PAH dioxygenase (*Mycobacterium sp.* S65) [24]	pdoA	[2Fe-2S]/Fe^2	unknown	unknown	unknown	AF546905
PAH dioxygenase (*Rhodococcus* strain I24) [21]	nidA	[2Fe-2S]/Fe^2	unknown	unknown	unknown	AF121905
PAH dioxygenase (*Pseudomonas abietaniphila* BKME-9) [60]	ditA1	[2Fe-2S]/Fe^2	[3Fe-4S]	unknown	unknown	AF119621
PAH dioxygenase (*N. aromaticivorans* F199) [37]	xylX	[2Fe-2S]/Fe^2	[2Fe-2S]	FAD	GR-type	AF079317
	bphA1a	[2Fe-2S]/Fe^2				
	bphA1b	[2Fe-2S]/Fe^2				
	bphA1c	[2Fe-2S]/Fe^2				
	bphA1d	[2Fe-2S]/Fe^2				
	bphA1e	[2Fe-2S]/Fe^2				
	bphA1f	[2Fe-2S]/Fe^2				
PAH dioxygenase (*Sphingomonas sp.* P2) [36]	ahdA1a	[2Fe-2S]/Fe^2	[2Fe-2S]	FAD	GR-type	AB091693
	ahdA1b	[2Fe-2S]/Fe^2				
	ahdA1c	[2Fe-2S]/Fe^2				
	ahdA1d	[2Fe-2S]/Fe^2				
	ahdA1e	[2Fe-2S]/Fe^2				
	xylX	[2Fe-2S]/Fe^2				
PAH dioxygenase (*Sphingomonas sp.* CHY-1) [42]	phnA1a	[2Fe-2S]/Fe^2	[2Fe-2S]	FAD	GR-type	AJ633551
	phnA1b	[2Fe-2S]/Fe^2				
PAH dioxygenase (*Cycloclasticus sp.* A5) [43]	phnA1	[2Fe-2S]/Fe^2	[2Fe-2S]	FAD	FNR$_N$-type	AB102786
Aniline dioxygenase (*D. acidovorans* 7N) [61]	orf7NC	[2Fe-2S]/Fe^2	None	FAD/[2Fe-2S]	FNR$_C$-type	AB177545
Phthalate dioxygenase (*P. putida*) [62]	pht3	[2Fe-2S]/Fe^2	None	FMN/[2Fe-2S]	FNR$_C$-type	D13229
Vanillate O-demethylase oxygenase (*Pseudomonas sp.* HR199) [63]	vanA	[2Fe-2S]/Fe^2	None	FMN/[2Fe-2S]	FNR$_C$-type	Y11521
Anthranilate 1,2-dioxygenase (*P. resinovorans* CA10) [64]	antA	[2Fe-2S]/Fe^2	None	FAD/[2Fe-2S]	FNR$_N$-type	AB047548
Naphthalene dioxygenase (*Ralstonia* sp. U2) [38]	nagAc	[2Fe-2S]/Fe^2	[2Fe-2S]	FAD/[2Fe-2S]	FNR$_N$-type	AF036940
Salicylate 5-hydroxylase (*Ralstonia* sp. U2) [65]	nagG	[2Fe-2S]/Fe^2	[2Fe-2S]	FAD/[2Fe-2S]	FNR$_N$-type	AF036940
Salicylate 5-hydroxylase (*P. aeruginosa* JB2) [66]	hybB	[2Fe-2S]/Fe^2	[2Fe-2S]	FAD/[2Fe-2S]	FNR$_N$-type	AF087482
Naphthalene dioxygenase (*P. aeruginosa* Pak1) [67]	pahA3	[2Fe-2S]/Fe^2	[2Fe-2S]	FAD/[2Fe-2S]	FNR$_N$-type	D84146
Carbazole 1,9a-dioxygenase (*Pseudomonas* sp. XLDN4-9)	carAa	[2Fe-2S]/Fe^2	[2Fe-2S]	FAD/[2Fe-2S]	FNR$_N$-type	DQ060076
PAH dioxygenase (*R. erythropolis* TA421) [68]	bphA1	[2Fe-2S]/Fe^2	[2Fe-2S]	FAD	GR-type	D88021
Chlorobenzene dioxygenase (*Pseudomonas* sp. P51) [17]	tcbAa	[2Fe-2S]/Fe^2	[2Fe-2S]	FAD	GR-type	U15298
Cumene dioxygenase (*P. fluorescens* IP01) [69]	cumA1	[2Fe-2S]/Fe^2	[2Fe-2S]	FAD	GR-type	D37828
p-Cumate dioxygenase (*R. palustris* No.7) [70]	psbAc	[2Fe-2S]/Fe^2	[2Fe-2S]	FAD	GR-type	AB022919
p-Cumate dioxygenase (*P. putida* F1) [71]	cmtAb	[2Fe-2S]/Fe^2	[2Fe-2S]	FAD	GR-type	U24215
Anthranilate 1,2-dioxygenase (*B. cepacoa* DBO1) [72]	andAc	[2Fe-2S]/Fe^2	[2Fe-2S]	FAD	GR-type	AY223539

Sequence Analysis

ClustalX [27] was used to obtain multiple alignments followed by distance calculations for each RHO component. Pairwise and multiple alignments were carried out with the default parameters. The weight matrix, Gonnet 250, was used to obtain the pairwise distance (PD) matrix. Alignments were visualized and examined using GENEDOC [28]. Phylogenetic trees were constructed by NJ method [29] from distance data and visualized and manipulated with TREEVIEW 1.6.6 [30]. Tree was also generated using iTOL: Interactive Tree Of Life, an online phylogenetic tree viewer and Tree Of Life resource [31]. The reliability of the trees obtained was evaluated by 1000 bootstrap replications. The domain structures for reductase components were analyzed using the Conserved Domain (CD) Database of NCBI [32], which compares primary sequences with all of the known domain structures in the databases.

Statistical Analysis

Matlab™ (The Mathworks, Inc.) was used to implement the proposed algorithm and to perform the simulation. The program under Matlab generates random subsets among 43 RHO samples, performs cluster analysis using the pairwise distance (PD) matrices and obtains the PD value which maximizes the classification accuracy. In the program, ClustalX [27] was used to obtain multiple alignments and PD matrices. The total running time spent for the simulation used in this paper is less than an hour and the matlab source code (.m) of the program is available upon request.

Results

Construction of the new classification scheme was accomplished as shown in the flowchart (Figure 1). Briefly, we first analyzed a total of 130 RHO enzymes in which 25 well-characterized RHO enzymes were selected as standards to test our hypothesis for the proposed classification system (Table 1). From the sequence analysis of ETC components of the standard RHO samples, we extracted the classification keys that reflect the phylogenetic relationship among ETC components, which turned into the grouping criteria in the following classifications. Next, oxygenases were phylogenetically classified with the classification keys obtained from ETC analyses. Finally we constructed the new systematic classification through statistical justification and tested its applicability with 38 RHO samples.

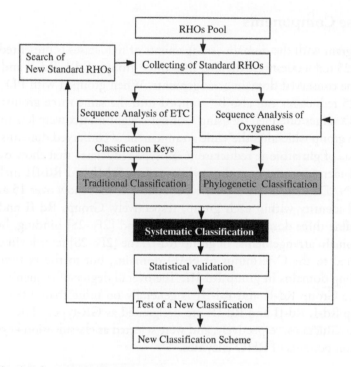

Figure 1. Flowchart for a new classification scheme.

Step 1: Sampling Standard RHOs

Table 1 shows the list of the 25 standard RHO samples whose genetic and functional information were completely available. It includes the members of Batie's three classes and Nam's four groups [16,18] and others that are not mentioned in those classifications. They are phenanthrene dioxygenase of Nocardioides sp. KP7 [11] and phthalate dioxygenases of Terrabacter sp. DBF63 [33], Arthrobacter keyseri 12B [23], M. vanbaalenii PYR-1 [10] and Rhodococcus sp. RHA1 [34]. The ferredoxin components of these RHO enzymes are the [3Fe-4S] type ferredoxin; only [2Fe-2S] type ferredoxin components were analyzed in the previous classification.

Step 2: Sequence Analysis of ETC Components from Standard RHO Samples

In step 2, phylogenetic information was obtained from the amino acid sequences of RHO ETC components, which in turn were converted into classification keys for grouping of oxygenase components for step 3 analyses.

Reductase Components

The phenogram with the domain arrangements of reductases is presented in Figure 2. The 25 reductases were divided into three groups, Rd-I, Rd-II and Rd-III, based on the conserved domain arrangements. When grouped with PD value of 0.85, the 25 reductases can also be grouped into the same three groups. Therefore, the PD values obtained by using Gonnet weight matrix were less than 0.85 within each group which has the same arrangement of conserved domains. Group Rd-I consists of glutathione reductase (GR) type reductases that show over 28% amino acid identity to one another, while groups Rd-II and Rd-III include the ferredoxin-NADP+ reductase (FNR) type reductases that show over 15 and 23% amino acid identity within each group, respectively. Groups Rd-II and Rd-III share the same three domains for flavin, NAD and [2Fe-2S] binding, but show different domain arrangements. In group Rd-II, the [2Fe-2S] ferredoxin domains are connected to the C-terminus of NAD domains, but to the N-terminus of flavin-binding domains in group Rd-III. The overall degree of sequence identity between the Group Rd-II and Rd-III is generally no more than 14%. Accordingly, group Rd-I, Rd-II and Rd-III are designated as GR-type, FNR_C-type and FNR_N-type reductases, respectively, and were selected as classification keys for the reductase components of RHO enzymes.

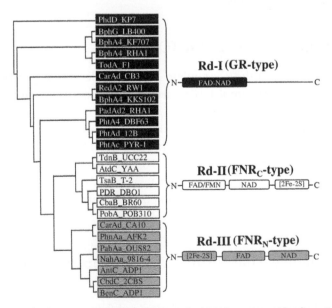

Figure 2. Grouping of reductase components from 25 standard RHO enzymes with schematic representation of the conserved domain structures. The names of bacterial strains are indicated after the enzyme names. GR-type, FNRC-type, and FNRN-type reductases are shown in the boxes with black, gray, and white background, respectively. Designations: FAD-Flavin adenine dinucleotide; NAD-Nicotinamide adenine dinucleotide.

Ferredoxin Components

Figure 3 shows the result of phylogenetic analysis for the ferredoxin components of standard RHO systems. Conserved amino acid residues were also shown to reveal the classification keys for ferredoxins. For the sequence alignment of ferredoxins, we initially aligned [2Fe-2S]-type and [3Fe-4S]-type ferredoxins separately and identified conserved amino acids. Multiple sequence alignment for the all ferredoxin sequences was then performed, from which we evaluated the validity of alignment. The tree shows that the 16 ferredoxins are divided into two distinct groups based on the type of iron-sulfur clusters, designated group Fd-I and Fd-II. The PD values within each group were less than 0.716. While the overall degree of amino acid sequence identity between groups is no more than 20%, members in each group show over 29% sequence identity to one another. In this analysis, Fdx3, a putidaredoxin-type ferredoxin isolated from Sphingomonas wittichii RW1, was used as an outgroup because Fdx3 is phylogenetically unrelated to those of RHO enzymes [35]. The group Fd-I comprises the [3Fe-4S] cluster-containing ferredoxins. Sequence alignments of this group revealed three conserved cysteine residues which serve as ligands for the [3Fe-4S] cluster. Group Fd-II consists of the [2Fe-2S] cluster-containing ferredoxins containing a highly conserved [2Fe-2S]-binding motif, $CXHX_nCX_2H$. Accordingly, group Fd-I and Fd-II are referred as [3Fe-4S]- and [2Fe-2S]-type ferredoxins, respectively, which in turn were selected as classification keys for ferredoxin components of RHOs.

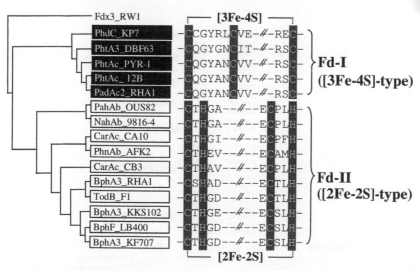

Figure 3. Grouping of ferredoxin components from standard RHO enzymes with the conserved sequence alignments. The amino acid residues involved in binding to [2Fe-2S] (CXHXnCXXH) and [3Fe-4S] (CX5CXnC) cluster are indicated by highlighted characters. Fdx3 was used as an outgroup. The names of bacterial strains are indicated after the enzyme names.

Step 3: Sequence Analysis and Grouping of Oxygenase Components from Standard RHO Samples

This step involves integration of the phylogenetic data of oxygenase components with respect to the classification keys obtained from ETC components. Figure 4 shows the dendrogram of 25 oxygenase components of standard RHO samples by the neighbor-joining (NJ) approach with CarAa from Pseudomonas resinovorans CA10 as an outgroup. The 25 oxygenase components can be clustered into two groups of homo (α_n) and hetero ($\alpha_n\beta_n$) oligomers (Figure 4). The homo-oligomer oxygenases include CbaA from Alcaligenes sp. BR60, PobA from Pseudomonas pseudoalcaligenes POB310, OphA2 from Burkholderia cepacia DBO1, TsaM from Comamonas testosteroni T-2 and CarAa from the strain CA10. However, when grouped with the classification keys from the phylogenetic analysis of ETC components, the 25 oxygenases are clustered into six distinct groups, designated Ox-I (an FNR_C-type reductase and a homo-oligomer oxygenase (α_n)), Ox-II (an FNR_C-type reductase and a hetero-oligomer type oxygenase ($\alpha_n\beta_n$)), Ox-III (an FNR_N-type reductase and an oxygenase ($\alpha_n\beta_n$)), Ox-IV (a hetero-oligomer oxygenase ($\alpha_n\beta_n$), a [2Fe-2S]-type ferredoxin and an FNR_N-type reductase), Ox-V (a hetero-oligomer oxygenase ($\alpha n\beta n$), a [2Fe-2S]-type ferredoxin and a GR-type reductase) and Ox-VI (a hetero-oligomer oxygenase ($\alpha_n\beta_n$), a [3Fe-4S]-type ferredoxin and a GR-type reductase) (Figure 4). Groups Ox-I, Ox-II and Ox-III consist of two-component RHO systems, while groups Ox-IV, Ox-V and Ox-VI include three-component RHO systems.

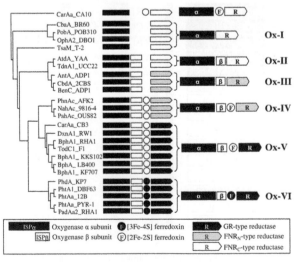

Figure 4. Phylogenetic clustering of oxygenase components from standard RHO enzymes with regard to the type of their ETC components. The tree was constructed by the NJ method with CarAa from P. resinovorans CA10 as an outgroup.

In Figure 5, the PD values within each group that shares the same classification keys were less than 0.61 with the exception of group Ox-I. This group can be further divided into four subgroups if using 0.61 as a PD value, where 25 oxygenases can be clustered into 10 groups including the outgroup of CarAa from CA10. Therefore, the PD value, 0.61, is a suitable criterion for grouping oxygenases based on our classification keys.

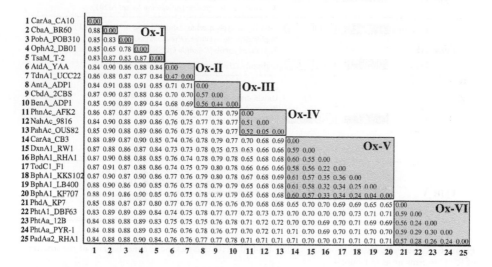

Figure 5. PD matrix of oxygenase components from standard RHO enzymes by Gonnet 250 weight matrix. The PD value for grouping in Figure 4 is 0.61. Each group is shown in shadow box with the exception of Ox-I. The group Ox-I can be further divided into four subgroups at the PD value of 0.61.

Step 4: Construction of the New Systematic Classification

In this step, the integrated classification data were converted to a systematic classification. As shown in Figure 6, the standard 25 RHO samples can be organized into 5 distinct types from 10 groups which clustered at the PD value of 0.61, designated type I, II, III, IV and V. All the members within each type share the same classification keys of ETC components. Type I and III can be further divided into two subtypes, being designated $\alpha\beta$ and α according to the type of oxygenase components in the RHO system; oxygenase components in the subtype $\alpha\beta$ and α are hetero-oligomers ($\alpha_n\beta_n$) and homo-oligomers (α_n), respectively.

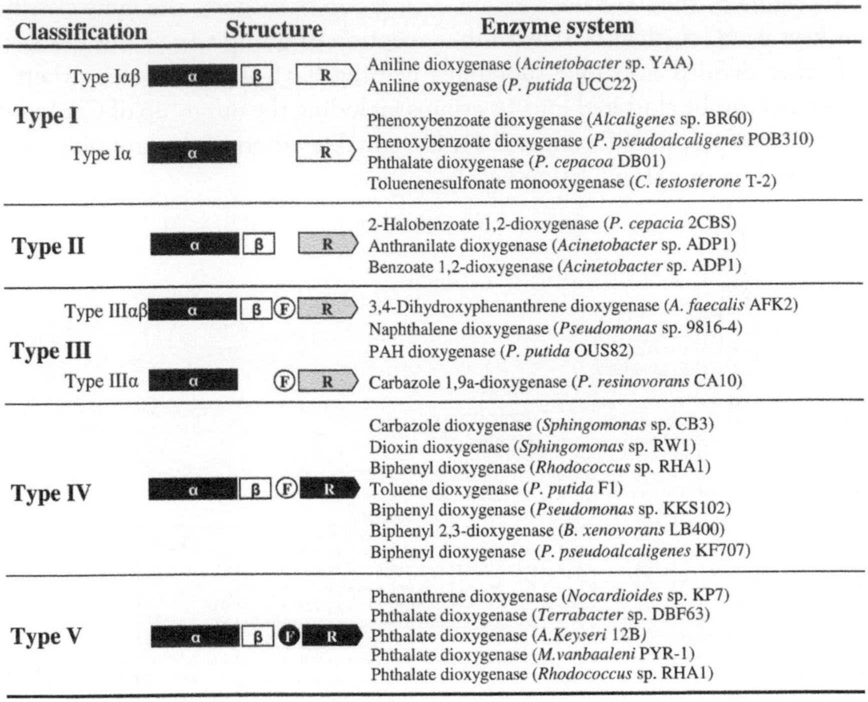

Classification	Structure	Enzyme system
Type I — Type Iαβ	α β R	Aniline dioxygenase (*Acinetobacter* sp. YAA) Aniline oxygenase (*P. putida* UCC22)
Type Iα	α R	Phenoxybenzoate dioxygenase (*Alcaligenes* sp. BR60) Phenoxybenzoate dioxygenase (*P. pseudoalcaligenes* POB310) Phthalate dioxygenase (*P. cepacoa* DB01) Toluenesulfonate monooxygenase (*C. testosterone* T-2)
Type II	α β R	2-Halobenzoate 1,2-dioxygenase (*P. cepacia* 2CBS) Anthranilate dioxygenase (*Acinetobacter* sp. ADP1) Benzoate 1,2-dioxygenase (*Acinetobacter* sp. ADP1)
Type III — Type IIIαβ	α β F R	3,4-Dihydroxyphenanthrene dioxygenase (*A. faecalis* AFK2) Naphthalene dioxygenase (*Pseudomonas* sp. 9816-4) PAH dioxygenase (*P. putida* OUS82)
Type IIIα	α F R	Carbazole 1,9a-dioxygenase (*P. resinovorans* CA10)
Type IV	α β F R	Carbazole dioxygenase (*Sphingomonas* sp. CB3) Dioxin dioxygenase (*Sphingomonas* sp. RW1) Biphenyl dioxygenase (*Rhodococcus* sp. RHA1) Toluene dioxygenase (*P. putida* F1) Biphenyl dioxygenase (*Pseudomonas* sp. KKS102) Biphenyl 2,3-dioxygenase (*B. xenovorans* LB400) Biphenyl dioxygenase (*P. pseudoalcaligenes* KF707)
Type V	α β F R	Phenanthrene dioxygenase (*Nocardioides* sp. KP7) Phthalate dioxygenase (*Terrabacter* sp. DBF63) Phthalate dioxygenase (*A.Keyseri* 12B) Phthalate dioxygenase (*M.vanbaaleni* PYR-1) Phthalate dioxygenase (*Rhodococcus* sp. RHA1)

Figure 6. A systematic classification scheme for standard RHO enzymes. Schematic diagram of the structures of RHO systems are as in Figure 4.

The type I system represents two-component RHOs that consist of an FNRC-type reductase and an oxygenase, which is either hetero-oligomer (type Iαβ) or homo-oligomer (type Iα). The RHO systems of group Ox-I and Ox-II belong to the type Iαβ and type Iα, respectively. Type II represents the other two-component systems that consist of an oxygenase and a FNRN-type reductase. The RHO enzymes in group Ox-III fall into type II. The well-characterized system for type II is benzoate 1,2-dioxygenase (BenABC) from Acinetobacter sp. ADP1. Type III represents three-component systems that consist of an oxygenase, a [2Fe-2S]-type ferredoxin and an FNRN-type reductase. All of group Ox-IV belong to type IIIαβ. The carbazole 1,9a-dioxygenase (CarAaAcAd) from P. resinovorans CA10 used as an outgroup for reconstructing the dendrogram of oxygenase components was determined to be type IIIα. The type IV systems represent another three-component systems that consist of an oxygenase, a [2Fe-2S]-type ferredoxin and a GR-type reductase. Type IV was shown to be the biggest group for the known RHO enzyme systems. Type V represents another different three-component systems that consist of an oxygenase, a [3Fe-4S]-type ferredoxin and a GR-type reductase. The RHO samples of the group Ox-VI were belonged to type V. The well-known example of the type V enzyme is phenanthrene dioxygenase (PhdAB-CD) from Nocardioides sp. KP7, in which the PhdC was the first [3Fe-4s]-type ferredoxin to be found in RHO enzyme systems.

Step 5: Statistical Justification of the New Systematic Classification

In this step, we statistically examined the classification scheme; the proposed PD value and the size and quality of the standard RHOs were evaluated using in-house program. Here, the learning accuracy is defined as the ratio of the number of correctly classified observations and the total number of observations.

First, we evaluated which PD value maximizes the classification accuracy for the 25 standard RHO set by the following algorithm:

INPUT: N oxygenases

(1) Select randomly four key oxygenases a_2, a_3, a_4 and a_5 from each categories II, III, IV, V respectively.

(2) At each of the N-1 steps p_0, p_1, ..., p_n, the closest two clusters are merged into a single cluster.

(3) Define cut-off distances as qi = $(p_{i-1} + p_i)/2$ for each i = 1,..., n. For each qi, a cluster is classified as category I, if there is no a_2-a_5 in the cluster. If one or more of a_i's (i = 2, ..., 5) are present in the cluster, the cluster is classified randomly as one of their category. Calculate the learning accuracy and find the cutoff which maximizes the learning accuracy.

Figure 7 is an output data showing the relations among learning accuracy, PD value and RHO classification result deduced from the 25 standard RHOs. This data indicates clearly that the PD value 0.611 satisfies demand for the classification of 25 standard RHO set with a maximum learning accuracy 1.0.

Based on the estimation of PD value, we devised a simulation to determine the suitable set of oxygenases which may minimize the classification error during the prediction. For the purpose of simulation, we used a total of 43 oxygenases which includes 25 standard and 18 test RHO samples. The training sets of 20, 25, 30, 35 and 40 RHOs were randomly selected from a total of 43 oxygenases for 50 times for each training set. Then, we evaluated the PD values which attain the maximum learning accuracy using the above algorithm for each set. Table 3 indicates that the mean deviation of such PD values is close to 0.61 and the values are not much affected by the number of observations, although mean PD values diminishes as the number of oxygenases increases. Figure 8 shows the simulation result based on 50 repetitions for 25 randomly selected RHOs, which indicates that the maximum accuracy is obtained at the mean PD value of 0.619. In Figure 9, it was further shown that the PD values of 25 oxygenases are most stable except for the set of entire 43 oxygenases, although there are some outliers which increase the variance. This simulation test can give us reliability of the 25 standard RHO set and PD value 0.611 for our classification.

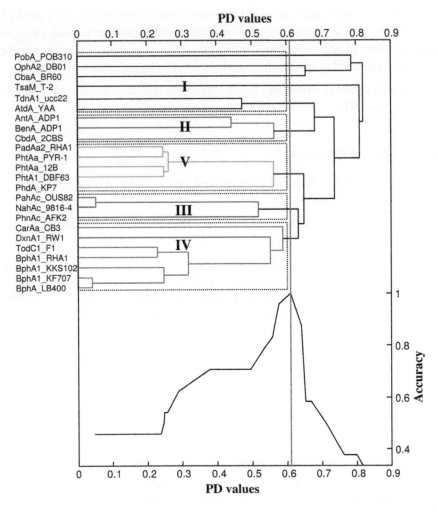

Figure 7. Accuracy plot and dendrogram of phylogenetic clustering of oxygenase components from standard RHO enzymes using the proposed classification algorithm. The right vertical axis represents the accuracy of classification based on clustering with the corresponding PD values on the horizontal axis. The labels along the left vertical axis display 25 standard RHO enzymes. At PD value of 0.61 (dashed vertical line), the standard enzymes have the maximum accuracy of 1.0.

Table 3. Mean and standard deviation of PD values for randomly selected 20, 25, 30, 35, 40, and 43 oxygenases based on 50 repetitions.

Number of oxygenases	Mean PD value	Standard deviation
20	0.6096	0.0191
25	0.6119	0.0146
30	0.6088	0.0117
35	0.6036	0.0137
40	0.6024	0.0096
43	0.5980	0.0105

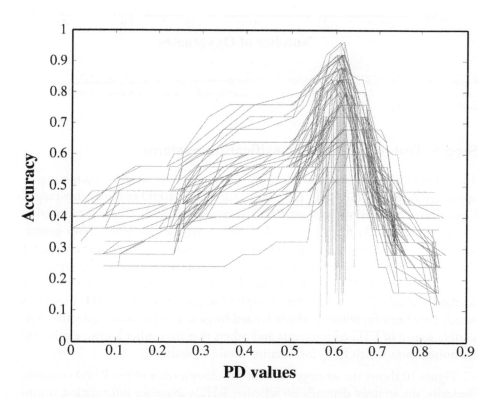

Figure 8. Accuracy plot using the proposed classification algorithm based on 50 repetitions of 25 randomly selected RHOs. The vertical axis represents the accuracy and the horizontal axis represents the PD values. Blue solid lines represent the accuracy of classification based on clustering with the corresponding PD values on the horizontal axis. Dashed vertical lines indicate the PD values at which the maximum accuracy is obtained for each repetition.

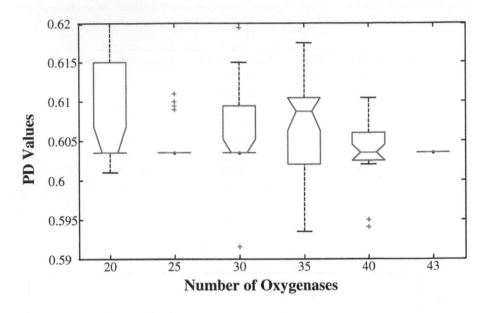

Figure 9. Box and whisker plot of PD values for randomly selected 20, 25, 30, 35, 40, and 43 oxygenases based on 50 repetitions. The box has lines at the lower quartile, median, and upper quartile values. Outliers are displayed with a + sign.

Step 6: Test of the New Classification Systems

The reliability of the new classification system was evaluated by examining its applicability and usefulness in the classification of 38 RHO enzymes that ranged broadly over the various type of RHO enzymes (Table 2). The selection criteria are as follows. At first, 14 complete RHO enzymes for which both genetic and functional information were well-known were selected. They were chosen for the purpose of verification of the classification systems, which are the same case as those standard RHO samples. Next, the 24 incomplete RHO enzymes were further selected. These incomplete samples include some of the RHO enzymes which have been functionally characterized by gene complementation with compatible source of ETC components and others that have other known equivalent homologs from which ETC information could be deduced.

Figure 10 shows the strategy used for the classification of test RHO enzymes. Basically, the strategy depends on whether RHOs sequence information is limited. In the case of complete RHO samples, query sequences were first subjected to multiple alignments followed by systematic classification; oxygenase components of RHOs undergo a phylogenetic classification whereas ETC sequences go through traditional classification route which are lastly consolidated into the final classification. In contrast, when analyzing incomplete RHOs, due to the lack of sequence information of the ETC component, only the oxygenase part can be classified according to the phylogenetic analysis. One of the merits of this classification strategy is that it responds dynamically to the growing pool of RHO enzymes. As shown in Figure 10, the output (final classification) of the classification system is returned to increase the standard sequence pool. As the standard RHO pool is growing, the classification gets more stable and objective and could possibly expand its coverage for many other RHO enzymes.

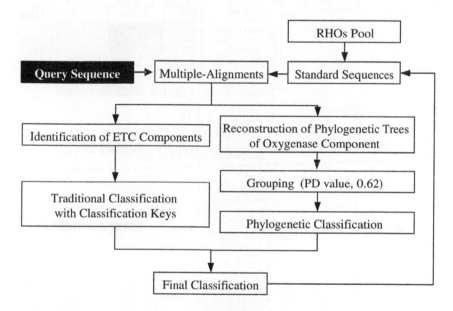

Figure 10. Strategy for query sequence classification used in this study.

Figure 11 shows the result of systematic/phylogenetic analysis for the classification of a total of 63 RHOs which comprised 25 standard and 38 test enzyme samples. This tree has a similar branching structure to that of the standard RHOs samples (Figure 3). It also mirrors well our classification scheme. When analyzing complete test RHO enzymes, the phylogenetic classification results of oxygenase components were in agreement with those of the traditional classification of ETC components. That is to say, ETC sequences of these complete RHO samples were identified by the classification keys to be the same types to which the respective oxygenase components were assigned. For example, both oxygenase (PahA3) and ETC components (PahA1A2) from P. aeruginosa Pak1 naphthalene dioxygenase were shown to be typed IIIA by phylogenetic and traditional classification, respectively. It indicates that our systematic classification reflects exactly the relationship/

partnership between oxygenase and ETC, which strongly demonstrates that our classification scheme has a potential to classify incomplete RHOs.

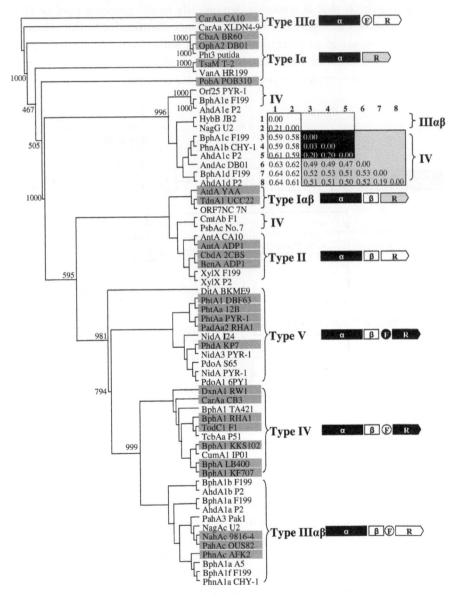

Figure 11. The resultant tree of the systematic/phylogenetic analysis for 63 RHO enzymes and PD matrix of the type IIIαβ group (HybB and NagG) and type IV group (BphA1c, PhnA1b, AhdA1c, BphA1d, and AhdA1d). The standard RHO enzymes are shown in shadow box. Schematic diagram of the structures of RHO systems are as in Figure 4. The tree was constructed by the NJ method with CarAa from P. resinovorans CA10 as an outgroup.

The 24 incomplete test RHO samples were all classified and assigned to each type by phylogenetic analysis. For instance, NidA and NidA3 from M. vanbaalenii PYR-1 were clustered together with phenanthrene dioxygenase (PhdABCD) from Nocardioides sp. KP7 being allocated to type V with a high bootstrap score and stable PD support. Oxygenases from Novosphingobium aromaticivorans F199 (BphA1a, BphA1b and BphA1f), Sphingomonas sp. P2 (AhdA1a and AhdA1b), Cycloclasticus sp. A5 (PhnA1) and Sphingomonas sp. CHY-1 (PhnA1a) were clustered into type IIIαβ with high bootstrap and stable PD values. In case of NagG from Ralstonia sp. U2, it was classified as type III with the complete RHO, HybB, from P. aeruginosa JB2. Another incomplete RHO samples, BphA1c and BphA1d from the strain F199, AhdA1c and AhdA1d from P2 and PhnA1b from CHY-1 were clustered as type IV together with the complete RHO, AndAc, from DBO1. Interestingly, BphA1c from the strain F199, AhdA1c from P2 and PhnA1b from CHY-1 of type IV can also be grouped into type IIIαβ with HybB from JB2 (2-hydroxybenzoate 5-salicylate hydroxylase) based on the PD value of 0.61 (Figure 8). This means that those three type IV RHOs are in the intersection with type IIIαβ. Both XylX oxygenase components from the strain F199 and P2 were grouped into the type II. Notably, ORF25, BphA1e and AhdA1e from the strain PYR-1, F199 and P2, respectively, formed a single distinctive separate cluster with a stable PD value; no standard or complete RHO is clustered together. In this case, although the genetic information of ETC for these three oxygenases are lacking, they were tentatively classified as type IV since BphA1e from the strain P2 was shown to be functionally active when complemented with the ETC component from type IV [36]. In fact, if all the classification keys are combined, we would be able to have 9 possible types for both two- and three-component systems. However, only 5 types were actually applied to classify RHO enzymes and we would need type X (alphabet) for the temporary classification of incomplete RHOs which do not have any standard RHO for comparison.

Step 7: Classification of a Total of 130 RHO Enzymes

As a final, a total of 130 RHO enzymes, which include 67 new RHO sequences, were classified using our classification system. Figure 12 shows an output based on the new classification system, which has a similar branching structure to that of the test set shown in step 6. Although RHO enzymes seem to have a grouping tendency according to their substrate specificities, the classification clearly indicates that the phylogenetic affiliations of RHO enzymes are determined mainly by their relationship with ETC components. For example, as shown in Type V, PhtAa (phthalate dioxygenase) and NidA3 (fluoranthene dioxygenase) from PYR-1 are grouped together. Although their substrates have been reported to be different, their ETC components have been experimentally shown to be the same as [3Fe-4S]-type ferredoxin and GR-type reductase [9,10,15,19]. In this respect, the results presented in Figure 12 well reflects the phylogenetic affiliations which are integrated with the interactions between RHO enzyme components.

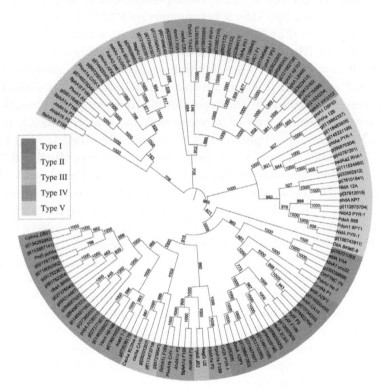

Figure 12. Classification of a total of 130 RHO enzymes. Some RHO sequences are provided with gi numbers whose information can be found in NCBI database.

Discussion

In many cases, RHO oxygenases have not been identified together with ETC components, thus it has been a challenging task to find a cognate ETC partner that can function in cooperation with these "incomplete" oxygenase components [9,35-39]. Since some ETCs are interchangeable with each other and can be often replaced by outside ETC sources and it is the terminal oxygenase component that has the catalytic active site, less emphasis was given to ETC components in Nam's classification. For example, DitA from P. abietaniphila BKME-9 and NidA from Rhodococcus sp. strain I24 were clustered in group III together with NahAc from Pseudomonas sp. NCIB9816-4 and PahAc from P. putida OUS82. However, these two oxygenases, DitA and NidA, which lacked ETC information (only ferredoxin for DitA is known), initially formed a group different from that of NahAc and PahAc in the phylogenetic classification route, being classified as type V that represents the three-component systems consisting of an oxygenase, a [3Fe-4S]-type ferredoxin and a GR-type reductase. Whereas the NahAc and PahAc were classified as type III which consists of an oxygenase, a [2Fe-2S]-type ferredoxin and an FNRN-type reductase. The discrepancy between these two classification's results comes from ETC information that reflects the relationship among oxygenases. The classification in Nam's scheme was based on pairwise sequence homology of oxygenase components, in which four percentage sets, 0–14, 15–24, 25–34 and 35–99%, were used. In the new classification, however, the grouping criterion for oxygenase components, which is a PD value of 0.62, was objectively determined based on ETC information. This PD value finally determines the two oxygenases, DitA and NidA, differently from the NahAc and PahAc. The results clearly show that information on oxygenases whose ETC components are not available can be used to draw ETC data in our classification system.

This new classification system provides insights into the evolutionary changes and relationships between oxygenase and ETC components. Seven oxygenase components, BphA1 [a-f] and XylX from N. aromaticivorans F199 [37] are of particular interest in this context. Although other ETC components are maybe involved, those oxygenases from F199, which were found to be scattered throughout several gene clusters, seem to function with a limited number of ETC components. Two ferredoxins (Rieske- [2Fe-2S]-type BphA3 and plant- [2Fe-2-S]-type XylT) and two reductases (GR-type BphA4 and FNR-type XylM) have been identified on the aromatic catabolic plasmid (pNL1) of F199. These oxygenases were classified as type II (XylX), IIIA (BphA1 [a,b,f]) and IV (BphA1 [c-e]) (Figure 8). Although there is no functional data available directly from the strain F199, genetic information with functional evidence have been reported from other sphingomonads which matches with those of F199. For example, sets of genes for an oxygenase from Sphingobium sp. P2 (AhdA1 [a-e]) [36], S. yanoikuyae B1 (BphA1c and XylX) [40,41] and Sphingobium sp. CHY-1 (PhnA1a and PhnA1b) [42] were reported to have the same gene arrangement with 63–97% sequence identity when compared to F199. Consequently, they were classified as the same types with those from F199; type II for XylX (B1), type IIIαβ for AhdA1 [a,b] (P2), PhnA1a (CHY-1) and type IV for BphA1c (B1), AhdA1 [c-e] (P2) and PhnA1b (CHY-1). We found that, without an exception, they were experimentally shown to use type IV ETC (all named as BphA3A4) that consists of [2Fe-2S]-type ferredoxin and GR-type reductase (Figure 6). These ETCs are highly homologous (> 79%) to the BphA3A4 from F199. Therefore, it is reasonable to predict that the 7 oxygenase components (BphA1 [a-f] and XylX) from F199 might be sharing the same type of ETC, BphA3A4. This implies that the limited ETC (BphA3A4) probably makes for strong selective pressure, favoring oxygenase components that can rapidly adapt by increasing their tolerance toward alternative ETC. This adaptation should be accompanied by the change of genetic information, which suggesting that RHO enzymes are probably evolving under the direction of selective pressure derived from ETCs. In fact, although BphA1c from F199 was clustered in type IV, it intersected with type III, the group of 2-hydroxybenzoate 5-salicylate hydroxylase (HybABCD) from P. aeruginosa JB2 (Figure 8). It suggests that BphA1c from F199 is probably evolving under the selective pressure derived from limited ETC. The XylX from the strain B1 is also worth noting, XylX belongs to type II, the two-component RHO enzymes, which normally functions with an FNR-type reductase. However, it was unexpectedly shown to be compatible with BphA3A4 [40,41]. Although both ferredoxins from type III and IV are [2Fe-2S]-type, the ferredoxin component (BphA3) from B1 seems phylogenetically inclined to be branched with ferredoxins which can complement oxygenases belong to type IIIαβ. The inclination of the ferredoxin component (BphA3) toward type IIIαβ also can be seen in the RHO enzyme from Cycloclasticus sp. A5, of which PhnA1 and PhnA3 components showed high identities over 62% with those (BphA1f and BphA3, respectively) from the strain F199 [43]. In this case, the oxygenase (PhnA1) belongs to type IIIαβ with the ETC components consist of a [2Fe-2S]-type ferredoxin (PhnA3) and an FNRN-type reductase (PhnA4). Taken together, the 7 oxygenases from F199, belonging to type II, IIIA and IV, were classified by our system to work with the three different ETC groups, FNRN-type reductase, [2Fe-2S]-type ferredoxins/FNRN-type reductase and [2Fe-2S]-type ferredoxins/GR-type reductase, respectively. However, all those oxygenases are assumed to function with [2Fe-2S]-type ferredoxin and GR-type reductase. From these observations question may arise about the degree of specificity in recognition between redox partners.

RHO pool is believed to respond dynamically to the environmental transitions such as substrate changes. In addition, relative tolerance between oxygenase and ETC components is also thought to be one of the important selective forces in evolution. In this context, even though little is known as to the role of ferredoxin component under selective pressure, it seems likely to be deeply involved in evolutionary changes. A possible benefit for using ferredoxin comes from its potential to promptly enhance the tolerance toward new redox partners. Cognizant of "the shorter, the faster and more effective", because of the relatively short sequence and simple structure of ferredoxin than that of reductase component, it has an evolutionary merit to adapt rapidly to dynamic environments. Hence, it might be an attractive hypothesis that ferredoxin has been evolutionarily chosen as a buffer between reductase and oxygenase component for rapid adaptation toward selective force. It might also be a strong driving force to affect the direction of evolution from two-component RHO to/toward three-component. Considering this two-way communication flow in RHO system, the adoption of ferredoxin as a new redox partner must have led to the changes in other partners, reductase and oxygenase, in RHO system. In this context, type III RHO seems likely to be a living fossil that has recorded evolutionary changes because they show transitional properties from two-component to three-component system. Two-component systems for type I and II share FNR-type reductase, FNRC- and FNRN-type, respectively, which consist of three domains (flavin, NAD and [2Fe-2S] binding). On the other hand, three-component systems for type IV and V, except for type III, share a GR-type reductase, which have no [2Fe-2S] cluster (Figure 2 and Figure 6). A [2Fe-2S] cluster of reductases separates the two-component system (type I and II) from the three-component system (type IV and V). It strongly demonstrates that the joining of ferredoxin to the two-component RHO system was probably the start of the type III three-component structure. During evolution, the type III FNRN-type reductases might have been gradually changed toward type IV and V GR-type reductases, although it is not sure whether it was recruited from other systems. In some cases, it might also be possible under extreme conditions that type I and II RHO systems evolved directly to type IV (or V) as seen in the XylX from S. yanoikuyae B1. Taken together, it may be postulated that RHOs are evolving under selective pressures from type I and II (two-component system) toward type IV and V (three-component), in which this change moves toward efficiency and keeps going continuously under dynamic environments.

Conclusion

In this study, we have developed a new classification system to classify bacterial Rieske non-heme iron aromatic ring-hydroxylating oxygenases. The new classification system presented in this study not only reflects sequence information but also interactions between RHO enzyme components. The system is characterized by the features that include the following. First, the new classification system analyzes RHO enzymes as a whole, in which information on each RHO components is organized into a system that is useful for the understanding of various aspects with respect to sequence, structure and biochemical function. Second, our new classification system is not static but responds dynamically to the growing pool of RHO enzymes. As standard RHO samples increase, the classification system evolves, which gets more stable and objective and could increase its coverage of many other RHO enzymes. Third, our classification can be applied reliably to the classification of incomplete RHOs. Fourth, the classification has direct applicability to experimental work. Our classification can provide the information about cognate ETC partners for oxygenase catalytic activity. Fifth, the system provides new insights into the evolution of RHO systems based on enzyme interaction.

Abbreviations

PAH, polycyclic aromatic hydrocarbons; RHO, Rieske-type non-heme iron aromatic ring-hydroxylating oxygenase; ETC, electron transport chain; PD, pairwise distance; NJ, neighbor-joing; CD, conserved domain; NR, nonredundant; GR, glutathione reductase; FNR, ferredoxin-NADP+ reductase.

Authors' Contributions

All authors participated in the design of the study and writing of the manuscript. CEC directed the whole research and critically revised the manuscript. All authors read and approved the final manuscript.

Acknowledgements

We thank H. Chen, CA Elkins and F. Rafii for critical review of on the manuscript. This work was supported in part by an appointment to the Postgraduate Research Program at the National Center for Toxicological Research administered by the Oak Ridge Institute for Science and Education through an interagency agreement between the U. S. Department of Energy and the U. S. Food and Drug Administration. The views presented in this article do not necessarily reflect those of the Food and Drug Administration

References

1. Sutherland JB, Rafii F, Khan AA, Cerniglia CE: Microbial transformation and degradation of toxic organic chemicals. In Mechanisms of polycyclic aromatic hydrocarbon degradation. Edited by: Young LY and Cerniglia CE. New York, Wiley-Liss Publication; 1995:269-306.

2. Cerniglia CE: Biodegradation of polycyclic aromatic hydrocarbons. Curr Opin Biotechnol 1993, 4:331-338.

3. Mason JR, Cammack R: The electron-transport proteins of hydroxylating bacterial dioxygenases. Annu Rev Microbiol 1992, 46:277-305.

4. Butler CS, Mason JR: Structure-function analysis of the bacterial aromatic ring-hydroxylating dioxygenases. Adv Microb Physiol 1997, 38:47-84.

5. Hayaishi O: History and scope. In Oxygenases. Edited by: Hayaishi O. New York, Academic Press; 1962:1–29.

6. Feig AL, Lippard SJ: Reactions of non-heme iron (II) centers with dioxygen in biology and chemistry. Chem Rev 1994, 94:124-133.

7. Raag R, Poulos TL: Crystal structure of the carbon monoxide-substrate-cytochrome P-450CAM ternary complex. Biochemistry 1989, 28:7586-7592.

8. Harayama S, Kok M, Neidle EL: Functional and evolutionary relationships among diverse oxygenases. Annu Rev Microbiol 1992, 46:565-601.

9. Stingley RL, Khan AA, Cerniglia CE: Molecular characterization of a phenanthrene degradation pathway in Mycobacterium vanbaalenii PYR-1. Biochem Biophys Res Commun 2004, 322:133-146.

10. Stingley RL, Brezna B, Khan AA, Cerniglia CE: Novel organization of genes in a phthalate degradation operon of Mycobacterium vanbaalenii PYR-1. Microbiology 2004, 150:3749-3761.

11. Saito A, Iwabuchi T, Harayama S: Characterization of genes for enzymes involved in the phenanthrene degradation in Nocardioides sp. KP7. Chemosphere 1999, 38:1331-1337.

12. Gibson DT, Resnick SM, Lee K, Brand JM, Torok DS, Wackett LP, Schocken MJ, Haigler BE: Desaturation, dioxygenation, and monooxygenation reactions catalyzed by naphthalene dioxygenase from Pseudomonas sp. strain 9816-4. J Bacteriol 1995, 177:2615-2621.

13. Chang HK, Zylstra GJ: Novel organization of the genes for phthalate degradation from Burkholderia cepacia DBO1. J Bacteriol 1998, 180:6529-6537.

14. Armengaud J, Happe B, Timmis KN: Genetic analysis of dioxin dioxygenase of Sphingomonas sp. strain RW1: catabolic genes dispersed on the genome. J Bacteriol 1998, 180:3954-3966.

15. Khan AA, Wang RF, Cao WW, Doerge DR, Wennerstrom D, Cerniglia CE: Molecular cloning, nucleotide sequence, and expression of genes encoding a polycyclic aromatic ring dioxygenase from Mycobacterium sp. strain PYR-1. Appl Environ Microbiol 2001, 67:3577-3585.

16. Batie CJ, Ballou DP, Correll CC: Phthalate dioxygenase reductase and related flavin-iron-sulfur containing electron transferases. In Chemistry and biochemistry of flavoenzymes. Volume III. Edited by: Müller F. Boca Raton, FL, CRC Press; 1992:543-556.

17. Werlen C, Kohler HP, van der Meer JR: The broad substrate chlorobenzene dioxygenase and cis-chlorobenzene dihydrodiol dehydrogenase of Pseudomonas sp. strain P51 are linked evolutionarily to the enzymes for benzene and toluene degradation. J Biol Chem 1996, 271:4009-4016.

18. Nam JW, Nojiri H, Yoshida T, Habe H, Yamane H, Omori T: New classification system for oxygenase components involved in ring-hydroxylating oxygenations. Biosci Biotechnol Biochem 2001, 65:254-263.

19. Kim SJ, Kweon OG, Freeman JP, Jones RC, Adjei MD, Jhoo JW, Edmondson RD, Cerniglia CE: Molecular cloning and expression of genes encoding a novel dioxygenase Involved in low- and high-molecular-weight polycyclic aromatic hydrocarbon degradation in Mycobacterium vanbaalenii PYR-1. Appl Environ Microbiol 2006, 72:1045-1054.

20. Saito A, Iwabuchi T, Harayama S: A novel phenanthrene dioxygenase from Nocardioides sp. strain KP7: expression in Escherichia coli. J Bacteriol 2000, 182:2134-2141.

21. Treadway SL, Yanagimachi KS, Lankenau E, Lessard PA, Stephanopoulos G, Sinskey AJ: Isolation and characterization of indene bioconversion genes from Rhodococcus strain I24. Appl Microbiol Biotechnol 1999, 51:786-793.

22. Nojiri H, Kamakura M, Urata M, Tanaka T, Chung JS, Takemura T, Yoshida T, Habe H, Omori T: Dioxin catabolic genes are dispersed on the Terrabacter sp. DBF63 genome. Biochem Biophys Res Commun 2002, 296:233-240.

23. Eaton RW: Plasmid-encoded phthalate catabolic pathway in Arthrobacter keyseri 12B. J Bacteriol 2001, 183:3689-3703.

24. Sho M, Hamel C, Greer CW: Two distinct gene clusters encode pyrene degradation in Mycobacterium sp. strain S65. FEMS Microbiol Ecol 2004, 48:209-220.

25. Krivobok S, Kuony S, Meyer C, Louwagie M, Willison JC, Jouanneau Y: Identification of pyrene-induced proteins in Mycobacterium sp. strain 6PY1: evidence for two ring-hydroxylating dioxygenases. J Bacteriol 2003, 185:3828-3841.

26. Altschul SF, Madden TL, Schaffer AA, Zhang J, Zhang Z, Miller W, Lipman DJ: Gapped BLAST and PSI-BLAST: a new generation of protein database search programs. Nucl Acids Res 1997, 25:3389-3402.

27. Jeanmougin F, Thompson JD, Gouy M, Higgins DG, Gibson TJ: Multiple sequence alignment with Clustal X. Trends Biochem Sci 1998, 23:403-405.

28. Nicholas KB, Nicholas HBJ, Deerfield DW: GeneDoc: Analysis and visualization of genetic variation. EMBNEW NEWS 1997, 4:14.

29. Saitou N, Nei M: The neighbor-joining method: a new method for reconstructing phylogenetic trees. Mol Biol Evol 1987, 4:406-425.

30. Page RD: TreeView: an application to display phylogenetic trees on personal computers. Comput Appl Biosci 1996, 12:357-358.

31. Letunic I, Bork P: Interactive Tree Of Life (iTOL): an online tool for phyloge-
    netic tree display and annotation. Bioinformatics 2007, 23:127-128.

32. Marchler-Bauer A, Anderson JB, DeWeese-Scott C, Fedorova ND, Geer LY,
    He S, Hurwitz DI, Jackson JD, Jacobs AR, Lanczycki CJ, Liebert CA, Liu C,
    Madej T, Marchler GH, Mazumder R, Nikolskaya AN, Panchenko AR, Rao
    BS, Shoemaker BA, Simonyan V, Song JS, Thiessen PA, Vasudevan S, Wang
    Y, Yamashita RA, Yin JJ, Bryant SH: CDD: a curated Entrez database of con-
    served domain alignments. Nucl Acids Res 2003, 31:383-387.

33. Habe H, Miyakoshi M, Chung J, Kasuga K, Yoshida T, Nojiri H, Omori T:
    Phthalate catabolic gene cluster is linked to the angular dioxygenase gene in
    Terrabacter sp. strain DBF63. Appl Microbiol Biotechnol 2003, 61:44-54.

34. Kitagawa W, Suzuki A, Hoaki T, Masai E, Fukuda M: Multiplicity of aromatic
    ring hydroxylation dioxygenase genes in a strong PCB degrader, Rhodococcus
    sp. strain RHA1 demonstrated by denaturing gradient gel electrophoresis. Bio-
    sci Biotechnol Biochem 2001, 65:1907-1911.

35. Armengaud J, Timmis KN: Molecular characterization of Fdx1, a putidaredox-
    in-type [2Fe-2S] ferredoxin able to transfer electrons to the dioxin dioxygenase
    of Sphingomonas sp. RW1. Eur J Biochem 1997, 247:833-842.

36. Pinyakong O, Habe H, Yoshida T, Nojiri H, Omori T: Identification of three
    novel salicylate 1-hydroxylases involved in the phenanthrene degradation of
    Sphingobium sp. strain P2. Biochem Biophys Res Commun 2003, 301:350-
    357.

37. Romine MF, Stillwell LC, Wong KK, Thurston SJ, Sisk EC, Sensen C, Gaas-
    terland T, Fredrickson JK, Saffer JD: Complete sequence of a 184-kilobase
    catabolic plasmid from Sphingomonas aromaticivorans F199. J Bacteriol 1999,
    181:1585-1602.

38. Jones RM, Britt-Compton B, Williams PA: The naphthalene catabolic (nag)
    genes of Ralstonia sp. strain U2 are an operon that is regulated by NagR, a
    LysR-type transcriptional regulator. J Bacteriol 2003, 185:5847-5853.

39. Armengaud J, Timmis KN: The reductase RedA2 of the multi-component di-
    oxin dioxygenase system of Sphingomonas sp. RW1 is related to class-I cyto-
    chrome P450-type reductases. Eur J Biochem 1998, 253:437-444.

40. Kim E, Zylstra GJ: Functional analysis of genes involved in biphenyl, naphtha-
    lene, phenanthrene, and m-xylene degradation by Sphingomonas yanoikuyae
    B1. J Ind Microbiol Biotechnol 1999, 23:294-302.

41. Cho O, Choi KY, Zylstra GJ, Kim YS, Kim SK, Lee JH, Sohn HY, Kwon GS,
    Kim YM, Kim E: Catabolic role of a three-component salicylate oxygenase from

Sphingomonas yanoikuyae B1 in polycyclic aromatic hydrocarbon degradation. Biochem Biophys Res Commun 2005, 327:656-662.

42. Demaneche S, Meyer C, Micoud J, Louwagie M, Willison JC, Jouanneau Y: Identification and functional analysis of two aromatic-ring-hydroxylating dioxygenases from a sphingomonas strain that degrades various polycyclic aromatic hydrocarbons. Appl Environ Microbiol 2004, 70:6714-6725.

43. Kasai Y, Shindo K, Harayama S, Misawa N: Molecular characterization and substrate preference of a polycyclic aromatic hydrocarbon dioxygenase from Cycloclasticus sp. strain A5. Appl Environ Microbiol 2003, 69:6688-6697.

44. Sato SI, Nam JW, Kasuga K, Nojiri H, Yamane H, Omori T: Identification and characterization of genes encoding carbazole 1,9a-dioxygenase in Pseudomonas sp. strain CA10. J Bacteriol 1997, 179:4850-4858.

45. Nakatsu CH, Wyndham RC: Cloning and expression of the transposable chlorobenzoate-3,4-dioxygenase genes of Alcaligenes sp. strain BR60. Appl Environ Microbiol 1993, 59:3625-3633.

46. Dehmel U, Engesser KH, Timmis KN, Dwyer DF: Cloning, nucleotide sequence, and expression of the gene encoding a novel dioxygenase involved in metabolism of carboxydiphenyl ethers in Pseudomonas pseudoalcaligenes POB310. Arch Microbiol 1995, 163:35-41.

47. Locher HH, Leisinger T, Cook AM: 4-Toluene sulfonate methyl-monooxygenase from Comamonas testosteroni T-2: purification and some properties of the oxygenase component. J Bacteriol 1991, 173:3741-3378.

48. Fujii T, Takeo M, Maeda Y: Plasmid-encoded genes specifying aniline oxidation from Acinetobacter sp. strain YAA. Microbiology 1997, 143:93-99.

49. Fukumori F, Saint CP: Nucleotide sequences and regulatory analysis of genes involved in conversion of aniline to catechol in Pseudomonas putida UCC22(pTDN1). J Bacteriol 1997, 179:399-408.

50. Haak B, Fetzner S, Lingens F: Cloning, nucleotide sequence, and expression of the plasmid-encoded genes for the two-component 2-halobenzoate 1,2-dioxygenase from Pseudomonas cepacia 2CBS. J Bacteriol 1995, 177:667-675.

51. Bundy BM, Campbell AL, Neidle EL: Similarities between the antABC-encoded anthranilate dioxygenase and the benABC-encoded benzoate dioxygenase of Acinetobacter sp. strain ADP1. J Bacteriol 1998, 180:4466-4474.

52. Kiyohara H, Nagao K, Kouno K, Yano K: Phenanthrene-degrading phenotype of Alcaligenes faecalis AFK2. Appl Environ Microbiol 1982, 43:458-461.

53. Takizawa N, Kaida N, Torigoe S, Moritani T, Sawada T, Satoh S, Kiyohara H: Identification and characterization of genes encoding polycyclic aromatic

hydrocarbon dioxygenase and polycyclic aromatic hydrocarbon dihydrodiol dehydrogenase in Pseudomonas putida OUS82. J Bacteriol 1994, 176:2444-2449.

54. Shepherd JM, Lloyd-Jones G: Novel carbazole degradation genes of Sphingomonas CB3: sequence analysis, transcription, and molecular ecology. Biochem Biophys Res Commun 1998, 247:129-135.

55. Masai E, Yamada A, Healy JM, Hatta T, Kimbara K, Fukuda M, Yano K: Characterization of biphenyl catabolic genes of gram-positive polychlorinated biphenyl degrader Rhodococcus sp. strain RHA1. Appl Environ Microbiol 1995, 61:2079-2085.

56. Wackett LP: Toluene dioxygenase from Pseudomonas putida F1. Methods Enzymol 1990, 188:39-45.

57. Ohtsubo Y, Nagata Y, Kimbara K, Takagi M, Ohta A: Expression of the bph genes involved in biphenyl/PCB degradation in Pseudomonas sp. KKS102 induced by the biphenyl degradation intermediate, 2-hydroxy-6-oxo-6-phenyl-hexa-2,4-dienoic acid. Gene 2000, 256:223-228.

58. Erickson BD, Mondello FJ: Nucleotide sequencing and transcriptional mapping of the genes encoding biphenyl dioxygenase, a multicomponent polychlorinated-biphenyl-degrading enzyme in Pseudomonas strain LB400. J Bacteriol 1992, 174:2903-2912.

59. Taira K, Hirose J, Hayashida S, Furukawa K: Analysis of bph operon from the polychlorinated biphenyl-degrading strain of Pseudomonas pseudoalcaligenes KF707. J Biol Chem 1992, 267:4844-4853.

60. Martin VJJ, Mohn WW: A novel aromatic-ring-hydroxylating dioxygenase from the diterpenoid-degrading bacterium Pseudomonas abietaniphila BKME-9. J Bacteriol 1999, 181:2675-2682.

61. Urata M, Uchida E, Nojiri H, Omori T, Obo R, Miyaura N, Ouchiyama N: Genes involved in aniline degradation by Delftia acidovorans strain 7N and its distribution in the natural environment. Biosci Biotechnol Biochem 2004, 68:2457-2465.

62. Nomura Y, Nakagawa M, Ogawa N, Harashima S, Oshima Y: Genes in PHT plasmid encoding the initial degradation pathway of phthalate in Pseudomonas putida. J Ferment Bioeng 1992, 74:333-344.

63. Priefert H, Rabenhorst J, Steinbuchel A: Molecular characterization of genes of Pseudomonas sp. strain HR199 involved in bioconversion of vanillin to protocatechuate. J Bacteriol 1997, 179:2595-2607.

64. Urata M, Miyakoshi M, Kai S, Maeda K, Habe H, Omori T, Yamane H, Nojiri H: Transcriptional regulation of the ant operon, encoding two-component

anthranilate 1,2-dioxygenase, on the carbazole-degradative plasmid pCAR1 of Pseudomonas resinovorans strain CA10. J Bacteriol 2004, 186:6815-6823.

65. Zhou NY, Al-Dulayymi J, Baird MS, Williams PA: Salicylate 5-hydroxylase from Ralstonia sp. strain U2: a monooxygenase with close relationships to and shared electron transport proteins with naphthalene dioxygenase. J Bacteriol 2002, 184:1547-1555.

66. Hickey WJ, Sabat G, Yuroff AS, Arment AR, Perez-Lesher J: Cloning, nucleotide sequencing, and functional analysis of a novel, mobile cluster of biodegradation genes from Pseudomonas aeruginosa strain JB2. Appl Environ Microbiol 2001, 67:4603-4609.

67. Takizawa N, Iida T, Sawada T, Yamauchi K, Wang YW, Fukuda M, Kiyohara H: Nucleotide sequences and characterization of genes encoding naphthalene upper pathway of pseudomonas aeruginosa PaK1 and Pseudomonas putida OUS82. J Biosci Bioeng 1999, 87:721-731.

68. Kosono S, Maeda M, Fuji F, Arai H, Kudo T: Three of the seven bphC genes of Rhodococcus erythropolis TA421, isolated from a termite ecosystem, are located on an indigenous plasmid associated with biphenyl degradation. Appl Environ Microbiol 1997, 63:3282-3285.

69. Habe H, Kasuga K, Nojiri H, Yamane H, Omori T: Analysis of cumene (iso-propylbenzene) degradation genes from Pseudomonas fluorescens IP01. Appl Environ Microbiol 1996, 62:4471-4477.

70. Puskas LG, Inui M, Kele Z, Yukawa H: Cloning of genes participating in aerobic biodegradation of p-cumate from Rhodopseudomonas palustris. DNA Seq 2000, 11:9-20.

71. Eaton RW: p-Cumate catabolic pathway in Pseudomonas putida Fl: cloning and characterization of DNA carrying the cmt operon. J Bacteriol 1996, 178:1351-1362.

72. Chang HK, Mohseni P, Zylstra GJ: Characterization and regulation of the genes for a novel anthranilate 1,2-dioxygenase from Burkholderia cepacia DBO1. J Bacteriol 2003, 185:5871-5881.

The Camp-HMGA1-RBP4 System: A Novel Biochemical Pathway for Modulating Glucose Homeostasis

Eusebio Chiefari, Francesco Paonessa, Stefania Iiritano,
Ilaria Le Pera, Dario Palmieri, Giuseppe Brunetti,
Angelo Lupo, Vittorio Colantuoni, Daniela Foti, Elio Gulletta,
Giovambattista De Sarro, Alfredo Fusco and Antonio Brunetti

ABSTRACT

Background

We previously showed that mice lacking the high mobility group A1 gene (Hmga1-knockout mice) developed a type 2-like diabetic phenotype, in which cell-surface insulin receptors were dramatically reduced (below 10% of those in the controls) in the major targets of insulin action, and glucose intolerance was associated with increased peripheral insulin sensitivity.

This particular phenotype supports the existence of compensatory mechanisms of insulin resistance that promote glucose uptake and disposal in peripheral tissues by either insulin-dependent or insulin-independent mechanisms. We explored the role of these mechanisms in the regulation of glucose homeostasis by studying the Hmga1-knockout mouse model. Also, the hypothesis that increased insulin sensitivity in Hmga1-deficient mice could be related to the deficit of an insulin resistance factor is discussed.

Results

We first show that HMGA1 is needed for basal and cAMP-induced retinol-binding protein 4 (RBP4) gene and protein expression in living cells of both human and mouse origin. Then, by employing the Hmga1-knockout mouse model, we provide evidence for the identification of a novel biochemical pathway involving HMGA1 and the RBP4, whose activation by the cAMP-signaling pathway may play an essential role for maintaining glucose metabolism homeostasis in vivo, in certain adverse metabolic conditions in which insulin action is precluded. In comparative studies of normal and mutant mice, glucagon administration caused a considerable upregulation of HMGA1 and RBP4 expression both at the mRNA and protein level in wild-type animals. Conversely, in Hmga1-knockout mice, basal and glucagon-mediated expression of RBP4 was severely attenuated and correlated inversely with increased Glut4 mRNA and protein abundance in skeletal muscle and fat, in which the activation state of the protein kinase Akt, an important downstream mediator of the metabolic effects of insulin on Glut4 translocation and carbohydrate metabolism, was simultaneously increased.

Conclusion

These results indicate that HMGA1 is an important modulator of RBP4 gene expression in vivo. Further, they provide evidence for the identification of a novel biochemical pathway involving the cAMP-HMGA1-RBP4 system, whose activation may play a role in glucose homeostasis in both rodents and humans. Elucidating these mechanisms has importance for both fundamental biology and therapeutic implications.

Background

Insulin resistance is a metabolic condition found relatively frequently among humans with chronic hyperinsulinemia and in experimental animal models with defective insulin signaling [1-3]. Recently, a link has been established between peripheral insulin sensitivity and the retinol (vitamin A) metabolism, and insulin resistance in rodents and humans has been linked to abnormalities of the vitamin A signaling pathway [4-6]. According to these studies, impaired glucose uptake in adipose tissue results in secondary systemic insulin resistance through release of the adipose-derived serum RBP4 [4,5]. However, it is unknown whether RBP4 effects on insulin sensitivity are vitamin A-dependent or vitamin A-independent. RBP4 (also called RBP) is mainly produced by the liver but also by adipocytes [7]. In plasma, retinol-RBP4 is found in an equimolar complex with transthyretin (TTR), which is a thyroid hormone transport protein that is synthesized in and secreted from the liver. This ternary complex prevents retinol-RBP4 excretion by the kidney [8]. By impairing insulin signaling in muscle, RBP4 inhibits glucose uptake and interferes with insulin-mediated suppression of glucose production in the liver, causing blood glucose levels to rise [4]. Conversely, mice lacking the RBP4 gene show increased insulin sensitivity, and normalizing increased RBP4 serum levels improves insulin resistance and glucose intolerance [4].

HMGA1 is a small basic protein that binds to adenine-thymine (A-T) rich regions of DNA and functions mainly as a specific cofactor for gene activation [9]. HMGA1 by itself has no intrinsic transcriptional activity; rather, it can trans-activate promoters through mechanisms that facilitate the assembly and stability of a multicomponent enhancer complex, the so-called enhanceosome, that drives gene transcription [9,10].

As part of an investigation into the molecular basis regulating the human insulin receptor gene, we previously showed that HMGA1 is required for proper insulin receptor gene transcription [11,12]. More recently, we showed that loss of HMGA1 expression, induced in mice by disrupting the HMGA1 gene, caused a type 2-like diabetic phenotype, in which, however, impaired glucose tolerance and overt diabetes coexisted with a condition of peripheral insulin hypersensitivity [13]. Concomitant insulin resistance and insulin hypersensitivity in peripheral tissues may paradoxically coexist as observed in livers of lipodystrophic and ob/ob mice [14], as well as in Cdk4 knockout mice with defective pancreatic beta cell development and blunted insulin secretion [15]. The hypothesis that the paradoxical insulin hypersensitivity of Hmga1-deficient mice could be due to a deficit, in these animals, of RBP4 is supported by our data. Herein, by employing the Hmga1-knockout mouse model, we provide compelling evidence for the identification of a novel biochemical pathway involving HMGA1 and RBP4, whose activation by the cAMP pathway may play an important role in maintaining glucose metabolism homeostasis in vivo, in both rodents and humans. The importance of HMGA1 in RBP4 gene transcription was substantiated in Hmga1-deficient mice, in which loss of HMGA1 expression considerably decreased RBP4 mRNA abundance and RBP4 protein production.

Results

RBP4 Gene Transcription is Induced by HMGA1 and cAMP

We first performed experiments to see whether HMGA1 had a role in activating the mouse RBP4 gene promoter at the transcriptional level. To test this possibility, HepG2 human hepatoma cells and mouse Hepa1 hepatoma cells were cotransfected transiently with mouse RBP4-Luc reporter plasmid plus increasing amounts of the HMGA1 expression vector. As shown in Figure 1, overexpression of HMGA1 considerably increased RBP4-Luc activity in both cell types and this effect occurred in a dose-dependent manner. Consistent with these results, RBP4 mRNA abundance was increased in cells overexpressing HMGA1 and was reduced in cells pretreated with siRNA targeting HMGA1 (Figure 1), indicating that activation of the RBP4 gene requires HMGA1. These data were substantiated by chromatin immunoprecipitation (ChIP) assay, showing that binding of HMGA1 to the endogenous RBP4 locus was increased in whole, intact HepG2 and Hepa1 cells naturally expressing HMGA1, and was decreased in cells exposed to siRNA against HMGA1 (Figure 1). Based on these results, in addition to previous observations indicating that cAMP, or agents which elevate intracellular cAMP, increase RBP transcript levels [16], we were interested to see whether a functional link could be established between cAMP, HMGA1, and RBP4. To this end, we first confirmed and extended the observation made by Jessen and Satre [16] that RBP is induced by cAMP in Hepa1 cells. As measured by Northern blot analysis of total RNA (Figure 2), RBP4 mRNA increased $\approx$ 5-fold over the basal level in Hepa1 cells treated with 0.5 mM 8-bromo cAMP (Br-cAMP), a standard concentration for c-AMP induction experiments [16]. As shown in Figure 2, RBP4 mRNA levels increased starting at 3 h, peaking at 24 h and then declining, suggesting a transient transcriptional stimulation. To establish whether HMGA1 was required for basal and cAMP-dependent RBP4 transcription, we transfected the HMGA1 expression vector in Hepa1 cells treated or not with Br-cAMP and RBP4 protein levels were analyzed 48 h later by Western blot. As shown in Figure 2, RBP4 protein expression was enhanced in cells overexpressing HMGA1 and even further in cells treated with cAMP, in which an increase in HMGA1 protein expression was simultaneously observed, suggesting that induction of RBP4 by cAMP may occur, at least in part, through activation of endogenous HMGA1 expression. This hypothesis was supported by the fact that RBP4 was reduced in cAMP treated cells in which endogenous levels of HMGA1 were specifically lowered by transfecting cells with HMGA1 antisense expression plasmid (Figure 2). However, further experiments are needed to fully explain the role of cAMP on HMGA1 expression. The functional significance of HMGA1 in RBP4 gene expression was confirmed in transient transcription assays in Hepa1 (and differentiated 3T3-L1, data not shown) cells, in which overexpression of HMGA1 caused an increase in both basal and cAMP-induced Luc activity from the mouse RBP-Luc reporter plasmid (Figure 3). This effect was substantiated in HEK-293 cAMP-responsive cells, a cell line ideally suited for studying the effects of HMGA1 on transcription since it does not express appreciable levels of this protein. As shown in Figure 3, in support of the role that HMGA1 plays in the context of RBP4 gene, the direct effect of cAMP was less effective in promoting RBP4 transcription in HEK-293 cells expressing low levels of HMGA1, becoming considerably higher in cells with forced expression of HMGA1.

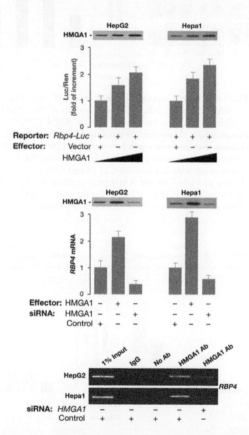

Figure 1. RBP4 gene expression is induced by HMGA1. (Top) Mouse RBP4-Luc reporter vector (2 μg) was transfected into HepG2 and Hepa1 cells plus increasing amounts (0, 0.5, or 1 μg) of HMGA1 expression plasmid. Data represent the means ± standard errors for three separate experiments; values are expressed as factors by which induced activity increased above the level of Luc activity obtained in transfections with RBP4-Luc reporter vector plus the empty expression vector, which is assigned an arbitrary value of 1. (Middle) HMGA1 expression plasmid was transfected into HepG2 and Hepa1 cells. After 6 h of transfection, the cells were treated with anti-HMGA1 (100 pmol), siRNA, or a non-targeting control siRNA, and endogenous RBP4 mRNA expression was measured 48 to 96 h later. Western blots of HMGA1 in each condition are shown in the autoradiograms. (Bottom) ChIP of the RBP4 promoter gene in HepG2 and Hepa1 cells, either untreated or pretreated with HMGA1 siRNA. ChIP was done using an anti-HMGA1 specific antibody (Ab).

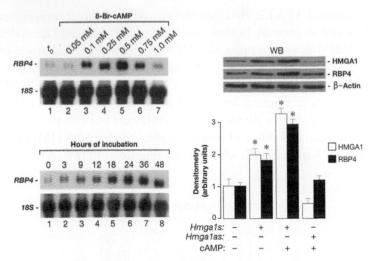

Figure 2. Stimulation of RBP4 mRNA and protein expression by cAMP and HMGA1. (Upper left) 20 μg of total RNA from Hepa1 cells treated with the indicated concentrations of Br-cAMP for 24 h (lanes 1–7) were analysed by Northern blot. Hybridization was carried out with an RBP4 cDNA or an 18S RNA probe as a control of the RNA loaded on each lane. (Lower left) 20 μg of total RNA from Hepa1 cells treated with 0.5 mM Br-cAMP for the indicated times were loaded on each lane (lanes 1–8) and analysed as above. (Right) Hepa1 cells, in the absence or presence of an expression plasmid (1 μg) containing the HMGA1 cDNA in either the sense (s) or antisense (as) orientation, were left untreated or treated with Br-cAMP (0.5 mM), total protein extracts were prepared 48 h later and HMGA1 and RBP4 protein expression levels were detected by Western blot (WB) with anti-HMGA1 and anti-RBP4 antibodies, respectively. β-actin, control of cellular protein loading. Densitometric analyses of three to five independent blots are shown.

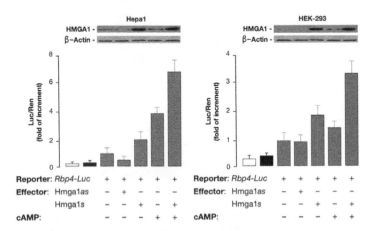

Figure 3. Role of HMGA1 in basal and cAMP-induced RBP4 expression. Rbp4-Luc reporter vector and HMGA1 expression plasmid (sense or antisense) were cotransfected into Hepa1 and HEK-293 cells, either untreated or treated with Br-cAMP. Data represent the means ± standard errors for three separate experiments. Transcriptional activity of the RBP4 gene promoter is shown as the ratio of luciferase activity to Renilla activity (Luc/Ren) as described in the experimental procedures. Values are expressed as the factors by which induced activity increased above the level of Luc activity obtained in transfections with the reporter vector alone, which is assigned an arbitrary value of 1. Open bar, mock (no DNA); black bar, pGL3-basic (vector without an insert). Western blots of HMGA1 and β-actin in each condition are shown in the autoradiograms.

Thus, these data together demonstrate that HMGA1 is of major importance for transcriptional regulation of the RBP4 gene, and indicate that a functional link exists between cAMP, HMGA1, and RBP4.

Hmga1-Deficient Mice Have Reduced Expression of RBP4 in Liver and Fat Tissue and Reduced Serum RBP4 Levels

In the light of the above experimental results, indicating that HMGA1 plays a positive role in RBP4 gene transcription in living cultured cells, it was interesting to analyze the functional consequences of genetic ablation of HMGA1 on RBP4 in vivo, in Hmga1-knockout mice. To this end, we performed studies aimed at investigating the expression of RBP4 mRNA and protein in Hmga1-deficient mice and wild-type controls. As shown in Figure 4, RBP4 mRNA was severely attenuated in both liver and fat from Hmga1-null mice, and reduced by 50% in Hmga1 heterozygous mutants, as assessed by real-time quantitative polymerase chain reaction (qRT-PCR). Reduced RBP4 mRNA levels in liver and adipose tissue paralleled the decrease in RBP4 serum levels as detected by Western blot analysis of serum samples from age- and body weight-matched mice with diverse genotypes (Figure 4), thereby showing the requirement of HMGA1 for full RBP4 expression in whole animals.

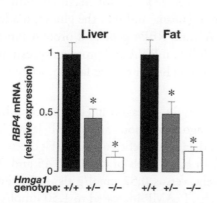 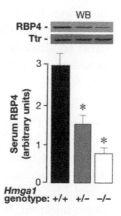

Figure 4. RBP4 expression in wild-type and Hmga1-deficient mice. RBP4 mRNA in liver and fat from control and Hmga1-deficient mice, as measured by qRT-PCR (left), and densitometric quantification of RBP4 serum levels as detected by Western blot (WB) of serum samples (2 μl) from mice with diverse genotypes (right). In WB analysis, an anti-transthyretin (TTR) antibody was used to confirm similar amounts of protein on each lane. Results are from 4–6 mice in each group. *P < 0.01 versus control mice.

HMGA1 and RBP4 Expression Increase in Liver and Fat of Normal Mice after Intraperitoneal Glucagon Injection

Based on our observations in intact cultured cells, indicating a role for the cAMP signaling pathway in HMGA1 and RBP4 gene expression, cAMP-inducible transcriptional activation of the Hmga1 and RBP4 genes was investigated in vivo, in whole animals, by systemic administration of the intracellular cAMP-elevating hormone glucagon. Under these conditions, glucagon-stimulated cAMP responses in terms of both Hmga1 and RBP4 mRNA expression were first analyzed in wild-type control mice. Consistent with our data in Hepa1 cells, Hmga1 and RBP4 mRNA levels significantly increased in liver and fat of normal mice after intraperitoneal injection of glucagon (Figure 5). Time course analyses revealed that the induction and accumulation of Hmga1 mRNA preceded the expression of RBP4 mRNA in both tissues. In liver, RBP4 mRNA appeared after that for Hmga1, peaked after 6 h following glucagon injection, and then remained at a plateau (Figure 5). In fat, RBP4 mRNA appeared at 1 h after Hmga1 mRNA, peaked after 6 h of glucagon stimulation, and decreased smoothly thereafter (Figure 5). Increased levels of Hmga1 and RBP4 mRNAs were paralleled by the increase of Hmga1 and RBP4 protein expression, as measured by Western blot analysis of proteins from liver and fat of glucagon-injected animals (Figure 5). Interestingly, when similar experiments were carried out in glucagon injected Hmga1-deficient mice, tissue expression of RBP4 mRNA was severely attenuated in liver and fat from heterozygous (Hmga1+/-) and Hmga1-null (Hmga1-/-) animals (Figure 6), thereby indicating that HMGA1 is indeed required for maximal induction of the RBP4 gene in vivo, in the whole organism, and that the glucagon/adenylate cyclase system regulates both HMGA1 and RBP4 gene and protein expression. Consistent with this conclusion, liver RBP4 mRNA and protein expression levels were lower in fed wild-type mice, becoming higher during fasting, when circulating glucagon increases (Figure 6).

As a measure of the glucagon efficacy in glucagon-injected mice, a liver biopsy was taken before and after glucagon injection, and cAMP levels in liver were determined for both control and Hmga1-deficient mice (Figure 6, inset). No substantial difference was found in basal levels of cAMP (0.45 and 0.50 in Hmga1-/- and Hmga1+/-, respectively, versus 0.52 µmol/g tissue in controls). After glucagon injection, hepatic levels of cAMP increased to 1.50 µmol/g tissue in control mice, compared with 1.48 and 1.52 in Hmga1-/- and Hmga1+/- mice, respectively. Results similar to those shown in the inset of Figure 6 were also obtained in epididymal and subcutaneous fat pads from control and mutant animals (data not shown), thus indicating that the glucagon-stimulated cAMP synthesis did not differ among mice with diverse genotypes.

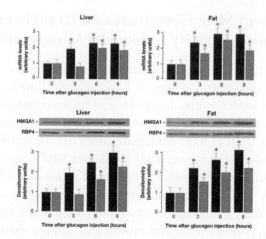

Figure 5. Hmga1 and RBP4 mRNA and protein expression in vivo, in glucagon-injected wild-type mice. Total RNA was isolated from liver (upper left) and fat (upper right) of 3-h-fasted mice, before and after intraperitoneal injection of glucagon. Levels of Hmga1 and RBP4 mRNA were measured at the indicated time intervals by qRT-PCR and normalized to RPS9 mRNA abundance, as described in Methods. Results are the mean values ± s.e.m. from 4–6 animals per group. Black bars, Hmga1 mRNA; gray bars, RBP4 mRNA. Representative Western blots from liver (lower left) and fat (lower right) of mice before and after glucagon injection are shown. Densitometric analyses of immunoblots are shown in bar graphs as the mean ± s.e.m. of data from 3–5 mice per each time point. Black bars, HMGA1; gray bars, RBP4.*P < 0.05 versus control mice (time 0).

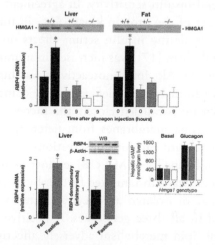

Figure 6. Comparison of RBP4 mRNA levels in glucagon-injected wild-type and Hmga1-deficient mice, and liver RBP4 expression in wild-type mice during fasting and fed. Total RNA was isolated from liver and fat of 3-h-fasted mice, before (time 0) and after 9 h of intraperitoneal injection of glucagon, and RBP4 mRNA was measured by qRT-PCR and normalized to RPS9 mRNA abundance. Results are the mean values ± s.e.m. from 6–8 animals per group. Black bars, Hmga1+/+, n = 8; gray bars, Hmga1+/-, n = 6; white bars, Hmga1-/-, n = 6. *P < 0.05 versus each control (time 0). Western blots for HMGA1 protein expression are shown in liver and fat from all three genotypes (top). The levels of RBP4 mRNA and protein (shown at the bottom of the figure) were measured in liver of fed and 6-h-fasted wild-type mice (6 animals per group), using qRT-PCR and Western blot (WB), respectively. *P < 0.05 versus fed mice. Inset, cAMP was measured in liver from control and Hmga1-deficient mice, in both basal conditions and 3 h after the intraperitoneal injection of glucagon (1 mg/kg body weight), as described in the Methods section. The data are mean ± s.e.m. for 4–6 animals per group.

Hmga1-Deficient Mice Have Increased Glut4 Expression and Insulin Signaling Activity In Skeletal Muscle and Fat

Systemic insulin resistance has been associated with elevation of serum RBP4, whereas genetic and pharmacological interventions aimed at decreasing serum RBP4 levels enhance insulin action and improve insulin sensitivity [4]. Increased peripheral insulin sensitivity during insulin-tolerance test was previously observed by us in Hmga1-knockout mice [13]. To verify whether a functional link indeed existed between HMGA1 and RBP4, and whether insulin hypersensitivity in Hmga1-deficient mice could be mediated by the HMGA1-RBP4 system, we carried out quantitative measurements of Glut4 mRNA transcript abundance. Examination by qRT-PCR showed a significant increase of Glut4 transcripts in both skeletal muscle and adipose tissues from Hmga1-deficient mice compared with controls (Figure 7). Accordingly, immunoblotting of muscle and fat tissue showed a 2- to 3-fold increase of Glut4 in the insulin hypersensitive Hmga1-knockout mice compared with controls (Figure 7), clearly indicating that an inverse correlation between RBP4 and Glut4 indeed exists in vivo, in this animal model of diabetes, in which reduced RBP4 may contribute to the maintenance of glucose homeostasis by increasing insulin signaling and peripheral insulin sensitivity. In agreement with this interpretation, the activation state of the protein kinase Akt, an important downstream target of PI 3-kinase regulating insulin serum effects on Glut4 translocation and carbohydrate metabolism [17], was increased in mutant animals. As shown in Figure 7, basal phospho-Akt immunoreactivity was higher in skeletal muscle and adipose tissues from Hmga1-deficient mice compared with wild-type controls, and this increase paralleled closely the increase of Glut4 protein in adipose and muscle plasma membranes from heterozygous and homozygous Hmga1 mutants. In line with previous observations on transcriptional repression of the mouse Glut4 gene by cAMP [18], endocrine upregulation of Glut4 in Hmga1-deficient mice was substantiated further by in vitro experiments (not shown), indicating that in isolated adipocytes treated with Br-cAMP, Glut4 mRNA was decreased in all three genotypes. A positive correlation of RBP4 levels with markers of lipid metabolism adversely affecting insulin sensitivity has been reported recently in both clinical and experimental studies [19,20]. Hmga1-knockout mice had lower levels of serum free fatty acids (0.45 ± 0.13 and 0.34 ± 0.07 in Hmga1+/+ and Hmga1-/-, respectively; $P < 0.05$), which might contribute to their improved insulin sensitivity.

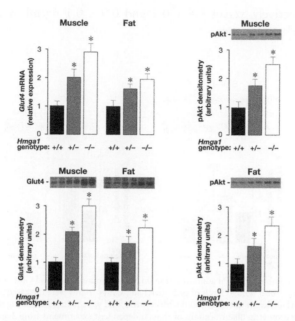

Figure 7. Glut4 and pAkt expression in wild-type and Hmga1-deficient mice. Glut4 mRNA (upper left) and protein content (lower left) in muscle and fat parallel pAkt protein abundance in skeletal muscle (upper right) and adipose tissue (lower right) from control and Hmga1-deficient mice. Representative Western blots of Glut4 and pAkt proteins are shown, together with the densitometric analyses of six to eight independent blots. Black bars, Hmga1[+/+], n = 8; gray bars, Hmga1[+/-], n = 6; white bars, Hmga1[-/-], n = 6. *P < 0.05 versus Hmga1[+/+].

Recombinant RBP4 Injection Reduces Glut4 and Insulin Signaling Activity in Muscle and Fat Tissue of Hmga1-Deficient Mice, and Attenuates Insulin Hypersensitivity of these Animals

To demonstrate that increased insulin sensitivity in mutant mice was directly due to HMGA1 regulation of RBP4, we determined the effect of recombinant RBP4 administration on Akt phosphorylation and Glut4 protein expression in skeletal muscle from Hmga1-deficient mice. As shown in Figure 8, Akt phosphorylation was reduced in muscle from RBP4-injected mutant animals compared with saline-injected Hmga1 mutants. The reduction in Akt phosphorylation in these genotypes correlated inversely with RBP4 serum levels in the same animals (Figure 8) and paralleled the reduction of Glut4 in skeletal muscle and adipose (not shown) plasma membranes (Figure 8), indicating that, in these conditions, activation of the Akt-Glut4 pathway is regulated, at least in part, by circulating RBP4. Plasma insulin levels were slightly higher in RBP4-injected mice, but no significant difference was found (1.6 ± 0.2 and 1.2 ± 0.2 in RBP4-injected Hmga1[+/-] and

Hmga1-/- respectively, versus 1.4 ± 0.1 and 0.9 ± 0.1 ng/ml in saline-injected Hmga1$^{+/-}$ and Hmga1$^{-/-}$ mice).

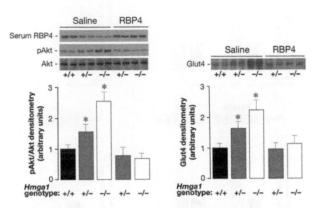

Figure 8. Effects of recombinant RBP4 administration on Akt phosphorylation and Glut4 protein expression in Hmga1-deficient mice. Basal (saline) levels of pAkt (left) and Glut4 (right) were increased in skeletal muscle of saline-injected Hmga1-deficient mice compared with controls, and were reduced following RBP4-injection (n = 6 per genotype). Densitometric quantifications of three independent experiments from 3 animals per genotype are shown, together with representative Western blots of pAkt, Glut4, and serum RBP4 of saline and RBP4-injected mice. *P < 0.05 versus Hmga1$^{+/+}$.

Consistent with the condition of insulin hypersensitivity, we previously reported that the glucose-lowering effect of exogenous insulin was enhanced in Hmga1-deficient mice during insulin-tolerance test (ITT) [13]. To support further the role of RBP4 in insulin hypersensitivity in Hmga1 mutants, we have determined the effect of RBP4 administration on the glucose fall induced by insulin in these genotypes during ITT. As shown in Figure 9, injection of human RBP4 in heterozygous and homozygous Hmga1 mutants caused a less dramatic fall in blood glucose levels, lessening the hypoglycemic response to intraperitoneal insulin observed in the saline-injected animals. Thus, taken together, our findings consistently support the role of HMGA1 as a key element in the transcriptional regulation of genes involved in glucose metabolism and add new insights into the compensatory mechanisms that may contribute to counteract insulin resistance in vivo. By directly regulating RBP4 gene transcription, HMGA1 enhances peripheral insulin sensitivity, ensuring glucose uptake in skeletal muscle. This, if on one hand might represent an adaptive mechanism to ameliorate insulin resistance in animals with a disadvantageous metabolic risk profile, on the other might indicate that the cAMP/HMGA1-mediated RBP4 expression during fasting (when glucagon peaks) may act physiologically to reduce insulin sensitivity in peripheral tissues, thereby contributing to the maintenance of euglycemia under this condition. This was supported by the observation that after an overnight fasting period

(12–16 h) plasma glucose concentration in wild-type mice was higher than that of Hmga1-deficient mice (89 ± 5 in Hmga1+/+ mice, versus 72 ± 6 and 62 ± 5 mg/dl in Hmga1+/- and Hmga1-/- mice, respectively; P < 0.05).

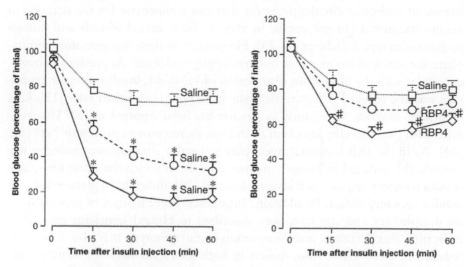

Figure 9. Effects of RBP4 on insulin sensitivity. Insulin-tolerance test (ITT) was assessed in Hmga1-deficient mice injected with saline alone (left), and in Hmga1 mutants injected chronically with purified RBP4 (right) (n = 6–8 per genotype in each condition). ITT was performed by measuring blood glucose levels in 12-h-fasted conscious mice injected intraperitoneally with human insulin (Human Actrapid, Novo Nordisk), 1 U/kg body weight. Open squares, Hmga1+/+; open circles, Hmga1+/-; open diamonds, Hmga1-/-. The degree of statistical significance was less in RBP4-injected Hmga1-deficient mice compared with the significance for saline-injected Hmga1 mutants. *P < 0.0001, saline-injected Hmga1-deficient mice versus Hmga1+/+; # P < 0.05, RBP4-injected Hmga1-/- mice versus Hmga1+/+.

Discussion

We have previously shown that loss of HMGA1 protein expression, induced in mice by disrupting the HMGA1 gene, severely decreased insulin receptor expression (below 10% of control animals) and phosphorylation in the major targets of insulin action, largely impaired insulin signaling, and reduced insulin secretion, producing a type 2-like diabetic phenotype in which defects in both peripheral insulin sensitivity and pancreatic beta-cell insulin secretion were coexpressed simultaneously [13]. However, despite the severe decrease in insulin receptor signaling and insulin receptor production, the glucose-lowering effect of exogenous insulin was enhanced in Hmga1-deficient mice during ITT, and the glucose infusion rate necessary to maintain euglycemia was higher in mutant mice during hyperinsulinemic-euglycemic clamp [13], supporting the existence of alternative pathways of insulin signaling promoting glucose uptake and disposal in certain adverse metabolic conditions such as those found in the Hmga1-knockout mouse. The existence of signaling pathways promoting glucose uptake and utilization in peripheral tissues through mechanisms that are independent of insulin has been postulated before, on the basis of experimental observations supporting the existence of molecular circuits/pathways that can compensate for the decrease in insulin-stimulated glucose uptake in vivo, in both animal models and human patients with type 2 diabetes [21-23]. However, how these compensatory mechanisms are activated has remained hitherto largely undefined. As previously shown, consistent with the ubiquitous distribution of HMGA1, insulin receptor expression was also reduced in pancreatic tissue from Hmga1-deficient mice [13]. Loss of insulin secretion in response to glucose has been reported in IRβ knockout mice with tissue-specific knockout of the insulin receptor in pancreatic beta cells [24]. As in the IRβ knockout mice, plasma insulin after glucose challenge was considerably reduced in Hmga1-mutant animals, in which the acute first-phase insulin secretory response was severely blunted [13], indicating a glucose-induced insulin secretory defect. In addition, substantial abnormalities in pancreatic islet morphology and size have been described in Hmga1-knockout mice [13], indicating that decreased insulin secretion in this genotype may also depend on reduced beta-cell mass. Thus, defects in both pancreatic beta-cell insulin secretion and peripheral insulin action coexist simultaneously in this knockout mouse model of diabetes, in which activation of compensatory mechanisms to efficiently overcome these metabolic abnormalities may be of vital importance.

Downregulation of Glut4 in adipose tissue is a typical feature of insulin-resistant states, such as obesity and type 2 diabetes [25]. It has been found that the decrease in Glut4 expression that occurs in the fatty tissue of obese animals and humans is accompanied by increased expression and secretion of the adipocyte-derived RBP4 fraction [4,5], suggesting that RBP4 production is tightly regulated by adipose tissue glucose uptake. RBP4 has been recently implicated in systemic insulin sensitivity in rodents and humans, in which elevated serum RBP4 levels were associated with reduced expression of Glut4 in adipocytes, and correlated inversely with peripheral insulin sensitivity. However, based on current data, the role of RBP4 in insulin sensitivity in humans is still controversial and might be restricted to rodent models only. Interspecies differences are known to exist and discrepancies between humans and mice might emphasize the role of non-genetic environmental factors and genetic modifiers in determining the phenotypic variations in RBP4 and insulin sensitivity between humans and animal models. Our results in the present study clearly indicate that in Hmga1-knockout mice RBP4 levels are considerably decreased in serum and in whole liver and adipose tissue extracts, strictly linking HMGA1 and RBP4 expression. We propose that HMGA1 deficiency adversely affects RBP4 expression and this, in animals with a disadvantageous metabolic risk profile like that observed in the Hmga1-knockout mouse model, might reflect an adaptive mechanism to increase glucose uptake and glucose disposal. Consistent with the results obtained in Hmga1-deficient mice, RBP4 was considerably reduced in cells of both human (HepG2) and mouse (Hepa1) origin readily expressing RBP4, following perturbation of endogenous HMGA1 protein expression in cells treated with siRNA against HMGA1. Conversely, an increase in RBP4 mRNA abundance was observed in both cell lines following forced expression of HMGA1, consistently supporting a role for HMGA1 in the transcriptional activation of the RBP4 gene. These findings were substantiated further by ChIP analysis, showing that HMGA1 indeed binds to the RBP4 locus in intact living cells.

Signal transduction pathways which raise intracellular cAMP have been reported to have a potential role in the regulation of RBP4 gene expression [16]. Although the molecular mechanisms underlying this effect remain poorly understood, evidence exists supporting the notion that the regulation of RBP4 gene transcription via the cAMP signaling pathway may be physiologically relevant. One important physiological condition in which intracellular cAMP increases is in response to low glucose availability. In this metabolic setting, a concomitant predominance of circulating counter-regulatory hormones, in particular pancreatic glucagon acting via the cAMP pathway, induces glycogenolysis and gluconeogenesis in the liver, which produce and release hepatic glucose in the blood. In this regard, the cAMP-element-binding protein (CREB) has been identified as a critical transcriptional checkpoint which, in response to cAMP, promotes hepatic glucose output through the synergistic activation of distinct transcriptional effector pathways, which include the PPAR gamma coactivator 1 (PGC1) and the NR4A orphan nuclear receptors [26].

In this paper, we report that systemic injection of glucagon to wild-type control mice caused an increase in RBP4 mRNA and protein expression, along with an increase of both intracellular cAMP and HMGA1 levels. Glucagon effects were attenuated in Hmga1-deficient mice, supporting a distinct role for HMGA1 in the regulation of RBP4 gene expression and functionally linking this two genes. As a consequence of the functional link between HMGA1 and RBP4, a significant increase in Glut4 mRNA and protein was observed in both skeletal muscle and adipose tissues from Hmga1-deficient mice compared with controls. An inverse relationship between RBP4 and Glut4 has been described previously, in the adipose-Glut4-/- mouse, in which the decrease in Glut4 expression that occurs in the fatty tissue of this mutant genotype is accompanied by increased expression and secretion of the fat-derived RBP4 [4]. In our model, instead, RBP4 expression is genetically impaired due to the lack of HMGA1 and Glut4 is increased in both muscle and fat, suggesting that abnormalities in RBP4 and/or metabolites of the vitamin A metabolism may directly affect whole-body insulin action and peripheral insulin sensitivity. In support of this possibility, identification of regulatory single nucleotide polymorphisms in the RBP4 gene associated with type 2 diabetes has been recently reported [27,28], while correlations of RBP4 with insulin resistance have been confirmed in experimental clinical approaches in humans [7]. Although conflicting results have been reported, raising doubt about the postulated relationship of RBP4 with insulin sensitivity in humans, our results in Hmga1-deficient mice confirm that an inverse correlation indeed exists between RBP4 and insulin sensitivity in vivo, in this animal model of diabetes, lending support to previous hypotheses that lowering RBP4 levels would be helpful in ameliorating insulin resistance, at least in mice.

Overall, our findings provide mechanistic insight into the regulation of glucose uptake and disposal in peripheral tissues, and support further the role of HMGA1 as a molecule that is likely to be an important emerging factor in the transcriptional activity of genes implicated in the maintenance of glucose homeostasis and metabolic control, such as the insulin receptor gene [11-13], the leptin gene [29], and, as shown here, the RBP4 gene. Apart from the intrinsic biological interest in elucidating the mechanisms leading to improvement in insulin sensitivity, a clear understanding of the molecular process involved is of potential importance in the development of new therapeutic strategies for patients with metabolic disorders such as obesity, diabetes, and other insulin resistant states.

Conclusion

We propose that HMGA1 can serve as a modulator of both RBP4 gene expression and protein function and represents an important novel mediator of glucose homeostasis in vivo.

Methods

Plasmids, Transfections, and ChIP

The RBP4-Luc reporter plasmid was obtained by cloning the NheI/XhoI 1427-bp sequence of the mouse RBP4 promoter (-1417 to +10) into pGL3 (Promega). This fragment was amplified from genomic DNA using the following modified primers: 5'-TTGCTAGCATGGCTAAGGTGCTTGTTGAAA-3', 5'-TTCTC-GAGCACACCCACTCCATCTCACC-3' and the integrity of this construct was checked by DNA sequencing. RBP4-Luc reporter plasmid, together with either the control vector plasmid or expression plasmid encoding HMGA1 [11], was transiently transfected into cultured cells using LipofectAMINE 2000 reagent

(Invitrogen), and Luc activity was assayed 48 h later, as previously described [30]. Renilla control vector served as an internal control of transfection efficiency, together with measurements of protein expression levels. For antisense HMGA1 experiments, RBP4-containing vector was cotransfected into Hepa1 cells with the expression plasmid pcDNA1 containing the HMGA1 cDNA in the antisense orientation [12]. Small interfering RNA (siRNA) targeted to HMGA1 [30] was transfected into cells at 50% to 60% confluency and cells were analyzed 48 to 96 h later. ChIP assay was performed in HepG2 and Hepa1 cells, either untreated or pretreated with HMGA1 siRNA as described previously [31]. Formaldehyde-fixed DNA-protein complex was immunoprecipitated with anti-HMGA1 antibody. Primers for the RBP4 sequence were used for PCR amplification of immunoprecipitated DNA (30 cycles), using PCR ready-to-go beads (Amersham Pharmacia Biotech). PCR products were electrophoretically resolved on 1.5% agarose gel and visualized by ethidium bromide staining.

Animals

Male Hmga1-deficient and wild-type mice aged 6–9 months were studied. The generation of these animals and many of the physiological characteristics of the mice have been described in detail [13]. All animal work was carried out at the Animal Facility at the 'Istituto dei Tumori di Napoli', and at the Faculty of Pharmacy, Roccelletta di Borgia, Catanzaro, using approved animal protocols and in accordance with institutional guidelines. Serum free fatty acid levels were measured in wild-type and Hmga1-knockout mice (n = 12–16 per genotype) using the NEFA C kit (Wako).

Real-Time PCR and Western Blot

For qRT-PCR, total cellular RNA was extracted from tissues using the RNAqueous-4PCR kit and subjected to DNase treatment (Ambion). RNA levels were normalized against 18S ribosomal RNA in each sample, and cDNAs were synthesized from 2 µg of total RNA using the RETROscript first strand synthesis kit (Ambion). Primers for mouse HMGA1 (NM_016660.2) (5'-GCAGGAAAAGGATGG-GACTG-3'; 5'-AGCAGGGCTTCCAGTCCCAG-3'), RBP4 (NM_011255.2) (5'-AGGAGAACTTCGACAAGGCT-3'; 5'-TTCCCAGTTGCTCAGAA-GAC-3'), Glut4 (NM_009204) (5'-TCATTGTCGGCATGGGTTT-3'; 5'-CGGCAAATAGAAGGAAGACGTA-3'), and RPS9 (NM_029767.2) (5'-CTG-GACGAGGGCAAGATGAAGC-3';5'-TGACGTTGGCGGATGAGCACA-3') were designed according to sequences from the GenBank database. A real-time thermocycler (Eppendorf Mastercycler ep realplex ES) was used to perform quantitative PCR. In a 20-μl final volume, 0.5 μl of the cDNA solution was mixed with SYBR Green RealMasterMix (Eppendorf), and 0.3 μM each of sense and antisense primers. The mixture was used as a template for the amplification by the following protocol: a denaturing step at 95°C for 2 min, then an amplification and quantification program repeated for 45 cycles of 95°C for 15 s, 55°C for 25 s, and 68°C for 25 s, followed by the melting curve step. SYBR Green fluorescence was measured, and relative quantification was made against the RPS9 cDNA used as an internal standard. All PCR reactions were done in triplicate.

Western blot analysis was performed to analyze HMGA1 and RBP4 protein expression in whole-cell liver and fat extracts from normal and mutant mice, using polyclonal specific antibodies raised against HMGA1 [11] and RBP4 (AdipoGen, Inc.). For the measurement of serum RBP4, blood was collected from the retro-orbital sinus, plasma protein extracts were resolved on 12% SDS-PAGE, blotted onto nitrocellulose membranes and RBP4 was detected using rabbit polyclonal antisera at 1:2000 dilution, as suggested by the manufacturer. TTR was detected using a goat anti-TTR polyclonal antibody (Santa Cruz Biotechnology). Rabbit anti-Glut4 polyclonal antibody was used as previously described [13].

In Vivo Studies with the Peptide Hormone Glucagon

For systemic administration of exogenous glucagon, mice were injected in the peritoneal cavity with human glucagon (1 mg/kg body weight) or saline after 3 h of fasting. At this dose, the peak increase of plasma glucagon in all genotypes was ~96% ± 10% above pre-injection levels, reflecting similar previous observations in rodents [32]. At different times after the injection the mice were killed by cervical dislocation, the liver and fat were rapidly removed, frozen into liquid nitrogen and stored at -80°C until processed. For cAMP determination, frozen samples were first homogenized in ice-cold trichloroacetic acid (TCA) (6% wt/vol), and cAMP was determined using the cAMP enzyme immunoassay kit (Amersham Pharmacia Biotech), according to the instructions specified by the manufacturer.

RBP4 Purification and Injection

Human RBP4 cDNA cloned into a pET3a expression vector was a kind gift from JW Kelly (The Scripps Research Institute). Based on previously published methodology [33], RBP4 protein expression vector was transformed into the BL21 strain of Escherichia coli (Stratagene), expanded in suspension culture and induced for 6 h with 1 mM isopropyl-D-thiogalactopyranoside to stimulate protein expression. Bacteria were pelleted and lysed by osmotic shock [34]. From this point on, all steps, including denaturation, refolding, and RBP4 purification, were performed essentially as described elsewhere [35]. Protein fractions were examined by sodium dodecyl sulfate polyacrylamide gel electrophoresis (SDS-PAGE) and immunoblotting, and desired fractions were pooled together, concentrated with an Amicon Centriprep-10 concentrator (Millipore), and stored at -80°C.

To determine whether elevation of RBP4 affected insulin hypersensitivity in vivo, in Hmga1-deficient mice, heterozygous and homozygous Hmga1 mutants, were intraperitoneally injected twice daily (at 12-h intervals) with 200 µg of purified human RBP4 (13 µg/g body weight per mouse) for 7 days. This resulted in a daily average serum level of human RBP4 similar to that of control mice (see Figure 8), which received physiological saline solution according to the same schedule above.

Statistical Analysis

The ANOVA test was used to evaluate the differences between the groups of mice. For all analyses, $P < 0.05$ was considered significant.

Abbreviations

Akt: protein kinase B; Br-cAMP: 8-bromo cAMP; cAMP: cyclic adenosine monophosphate; CREB: cAMP-element-binding protein; Glut4: glucose transporter-4; HEK-293: human embryonic kidney-293; Hepa1: mouse hepatoma; HMGA1: high mobility group A1; ITT: insulin-tolerance test; PGC1: PPAR gamma coactivator 1; qRT-PCR: quantitative Real-Time PCR; RBP4: retinol-binding protein 4; siRNA: small interfering RNA; TCA: trichloroacetic acid; TTR: transthyretin.

Authors' Contributions

EC and FP performed qRT-PCR studies as well as transient transfections with reporter and expression vectors, and participated in the design of the study. SI was involved in Western blotting studies and assisted FP in performing transfections with siRNA. ILP participated in Western blotting and performed cloning studies. DP, EG, GDS and AF participated in the analysis and discussion of the in vivo data from normal and mutant mice. GB, AL, and VC performed certain aspects of the assays detailed in Figures 1, 2 and 3 and contributed with Northern blotting studies. DF provided helpful discussion on this manuscript and participated in ChIP analysis. AB conceived, coordinated, and supervised the project, analyzed data, and wrote the manuscript. All authors read and approved the final manuscript.

Acknowledgements

We remain extraordinarily indebted to Drs R Citraro and N Costa and the entire staff of the animal facilities for animal care. We are most grateful to Dr JW Kelly for his generous gift of RBP4 expression vector, pET3a. We would also like to thank Mrs A Malta and Dr G Ceravolo for secretarial help. This work was supported by Telethon-Italy, grant GGP04245, and MIUR, protocol 2004062059-002 Italy (AB).

References

1. Kahn CR: Insulin action, diabetogenes, and the cause of type II diabetes (Banting Lecture). Diabetes 1994, 43:1066–1084.

2. Polonsky KS, Sturis J, Bell GI: Non-insulin-dependent diabetes mellitus – a genetically programmed failure of the beta cell to compensate for insulin resistance. New Engl J Med 1996, 334:777–783.

3. Taylor S: Insulin resistance or insulin deficiency: which is the primary cause of NIDDM? Diabetes 1994, 43:735–740.

4. Yang Q, Grahm TE, Mody N, Preitner F, Peroni OD, Zabolotny JM, Kotani K, Quadro L, Kahn BB: Serum retinol binding protein 4 contributes to insulin resistance in obesity and type 2 diabetes. Nature 2005, 436:356–362.

5. Graham TE, Yang Q, Blüher M, Hammarstedt A, Ciaraldi TP, Henry RR, Wason CJ, Oberbach A, Jansson P-A, Smith U, Kahn BB: Retinol-binding protein 4 and insulin resistance in lean, obese, and diabetic subjects. New Engl J Med 2006, 354:2552–2563.

6. Ziouzenkova O, Orasanu G, Sharlach M, Akiyama TE, Berger JP, Viereck J, Hamilton JA, Tang G, Dolnikowski GG, Vogel S, Duester G, Plutzky J: Retinaldehyde represses adipogenesis and diet-induced obesity. Nat Med 2007, 13:695–702.

7. von Eynatten M, Humpert PM: Retinol-binding protein-4 in experimental and clinical metabolic disease. Expert Rev Mol Diagn 2008, 8:289–299.

8. Monaco HL, Rizzi M, Coda A: Structure of a complex of two plasma proteins: transthyretin and retinol binding protein. Science 1995, 268:1039–1041.

9. Bustin M, Reeves R: High-mobility group proteins: architectural components that facilitate chromatin function. Prog Nucleic Acid Res Mol Biol 1996, 54:35–100.

10. Thanos D, Maniatis T: Virus induction of human IFN beta gene expression requires the assembly of an enhanceosome. Cell 1995, 83:1091–1100.

11. Brunetti A, Manfioletti G, Chiefari E, Goldfine ID, Foti D: Transcriptional regulation of human insulin receptor gene by the high-mobility group protein HMGI(Y). FASEB J 2001, 15:492–500.

12. Foti D, Iuliano R, Chiefari E, Brunetti A: A nucleoprotein complex containing Sp1, C/EBP beta, and HMGI-Y controls human insulin receptor gene transcription. Mol Cell Biol 2003, 23:2720–2732.

13. Foti D, Chiefari E, Fedele M, Iuliano R, Brunetti L, Paonessa F, Barbetti F, Croce CM, Fusco A, Brunetti A: Lack of the architectural factor HMGA1 causes insulin resistance and diabetes in humans and mice. Nat Med 2005, 11:765–773.

14. Shimomura I, Matsuda M, Hammer RE, Bashmakov Y, Brown MS, Goldstein JL: Decreased IRS-2 and increased SREBP-1c lead to mixed insulin resistance and sensitivity in livers of lipodystrophic and ob/ob mice. Mol Cell 2000, 6(1):77–86.

15. Rane SG, Dubus P, Mettus RV, Galbreath EJ, Boden G, Reddy EP, Barbacid M: Loss of Cdk4 expression causes insulin-deficient diabetes and Cdk4 activation results in β-islet cell hyperplasia. Nat Genet 1999, 22:44–52.

16. Jessen KA, Satre MA: Induction of mouse retinol binding protein gene expression by cyclic AMP in Hepa 1–6 cells. Arch Biochem Biophys 1998, 357: 126–130.

17. Cho H, Mu J, Kim JK, Thorvaldsen JL, Chu Q, Crenshaw EB 3rd, Kaestner KH, Bartolomei MS, Shulman GI, Birnbaum MJ: Insulin resistance and a diabetes mellitus-like syndrome in mice lacking the protein kinase Akt2 (PKB beta). Science 2001, 292:1728–1731.

18. Kaestner KH, Flores-Riveros JR, McLenithan JC, Janicot M, Lane MD: Transcriptional repression of the mouse insulin-responsive glucose transporter (GLUT4) gene by cAMP. Proc Natl Acad Sci USA 1991, 88:1933–1937.

19. von Eynatten M, Lepper PM, Liu D, Lang K, Baumann M, Nawroth PP, Bierhaus A, Dugi KA, Heemann U, Allolio B, Humpert PM: Retinol-binding protein 4 is associated with components of the metabolic syndrome, but not with insulin resistance, in men with type 2 diabetes or coronary artery disease. Diabetologia 2007, 50:1930–1937.

20. Stefan N, Hennige AM, Staiger H, Machann J, Schick F, Schleicher E, Fritsche A, Haring HU: High circulating retinol-binding protein 4 is associated with elevated liver fat but not with total, subcutaneous, visceral, or intramyocellular fat in humans. Diabetes Care 2007, 30:1173–1178.

21. Saltiel AR, Pessin JE: Insulin signaling pathways in time and space. Trends Cell Biol 2002, 12:65–71.

22. Kitamura T, Kahn CR, Accili D: Insulin receptor knockout mice. Annu Rev Physiol 2003, 65:313–332.

23. Bouché C, Serdy S, Kahn CR, Goldfine AB: The cellular fate of glucose and its relevance in type 2 diabetes. Endocr Rev 2004, 25(5):807–30.

24. Kulkarni RN, Bruning JC, Winnay JN, Postic C, Magnuson MA, Kahn CR: Tissue-specific knockout of the insulin receptor in pancreatic β cells creates an insulin secretory defect similar to that in type 2 diabetes. Cell 1999, 96: 329–339.

25. Shepherd PR, Kahn BB: Glucose transporters and insulin action-implications for insulin resistance and diabetes mellitus. New Engl J Med 1999, 341: 248–257.

26. Desvergne B, Michalik L, Wahli W: Transcriptional regulation of metabolism. Physiol Rev 2006, 86:465–514.

27. Munkhtulga L, Nakayama K, Utsumi N, Yanagisawa Y, Gotoh T, Omi T, Kumada M, Erdenebulgan B, Zolzaya K, Lkhagvasuren T, Iwamoto S: Identification of a regulatory SNP in the retinol binding protein 4 gene associated with type 2 diabetes in Mongolia. Hum Genet 2007, 120:879–888.

28. Craig RL, Chu WS, Elbein SC: Retinol binding protein 4 as a candidate gene for type 2 diabetes and prediabetic intermediate traits. Mol Genet Metab 2007, 90:338–344.

29. Melillo RM, Pierantoni GM, Scala S, Battista B, Fedele M, Stella A, De Biasio MC, Chiappetta G, Fidanza V, Condorelli G, Santoro M, Croce CM, Viglietto G, Fusco A: Critical role of the HMGI(Y) proteins in adipocytic cell growth and differentiation. Mol Cell Biol 2001, 21:2485–2495.

30. Paonessa F, Foti D, Costa V, Chiefari E, Brunetti G, Leone F, Luciano F, Wu F, Lee AS, Gulletta E, Fusco A, Brunetti A: Activator protein-2 overexpression accounts for increased insulin receptor expression in human breast cancer. Cancer Res 2006, 66:5085–5093.

31. Costa V, Paonessa F, Chiefari E, Palaia L, Brunetti G, Gulletta E, Fusco A, Brunetti A: The insulin receptor: a new anticancer target for peroxisome proliferator-activated receptor-g (PPARg) and thiazolidinedione-PPARg agonists. Endocr Relat Cancer 2008, 15:325–335.

32. Velliquette RA, Koletsky RJ, Ernsberger P: Plasma glucagon and free fatty acid responses to a glucose load in the obese spontaneous hypertensive rat (SHROB) model of matabolic syndrome X. Exp Biol Med (Maywood) 2002, 227:164–170.

33. Xie Y, Lashuel HA, Miroy GJ, Dikler S, Kelly JW: Recombinant human retin-ol-binding protein refolding, native disulfide formation, and characterization. Protein Expr Purif 1998, 14:31–37.

34. Burger A, Berendes R, Voges D, Huber R, Demange P: A rapid and efficient purification method for recombinant annexin V for biophysical studies. FEBS Lett 1993, 329:25–28.

35. Isken A, Golczak M, Oberhauser V, Hunzelmann S, Driever W, Imanishi Y, Palczewski K, von Lintig J: RBP4 disrupts vitamin A uptake homeostasis in a STRA6-deficient animal model for Matthew-Wood syndrome. Cell Metab 2008, 7:258–268.

Advances in Parasite Genomics: From Sequences to Regulatory Networks

Elizabeth A. Winzeler

ABSTRACT

Parasites have kept many secrets from the researchers who have sought to erad-icate them over past decades. The mechanisms by which they evade drugs, es-cape the immune system, regulate switching between genes involved in im-mune evasion, and orchestrate development have been difficult to elucidate. They have been successful at this in part because they are difficult to keep in the laboratory, difficult to breed, and difficult to raise in sufficient quanti-ties for biochemistry, and because they parasitize hosts that are not ideal ex-perimental subjects. While Plasmodium falciparum is less tractable than one would wish, genetic manipulation can still be performed. On the other hand, Plasmodium vivax, which cannot be maintained in culture, is even less ac-cessible, and there are few research tools available.

While these difficulties present impediments to drug, vaccine, and basic research, the availability of parasite genome sequences and related genome-based tools have provided substantial opportunities to overcome the lack of a robust culture system needed for traditional molecular biology, the shortage of material for biochemistry, and the lack of traditional genetic methods for studying gene function. The advent of new technologies for examining and detecting genetic variation, measuring transcript abundance, and measuring protein or metabolite abundance on a genome-wide scale, or for sequencing genomes in combination with new computational methods, may lift some of the barriers to working on actual pathogens. Here, I will review some recent discoveries that were facilitated by industrial-scale molecular biology approaches.

New Genome Sequences

The year 2008 witnessed the publication of the complete genome sequence of P. vivax as well as that of Plasmodium knowlesi [1],[2]. Although P. vivax may be responsible for up to 40% of the 515 million malaria cases each year, work on this parasite has generally lagged because it cannot yet be maintained in long-term culture. Among the highlights of the P. vivax genome sequence was the observation that it encodes a variety of cell-binding proteins involved in erythrocyte selection, and thus P. vivax may be able to use a variety of red cell invasion strategies. Of course, knowing the complement of genes encoded by a genome only serves as a prelude to further functional studies, and the first set of gene expression data for P. vivax was published soon afterwards [3]. This work showed that the transcriptional program of P. vivax is similar to that of P. falciparum and offers hypotheses about the function of a variety of P. vivax genes. For example, a gene whose transcriptional pattern is correlated with those of known invasion genes may also be involved in invasion. Accompanying the publication on the P. vivax genome was the sequence of P. knowlesi, described as the fifth human pathogen given its documented zoonoses [4]. This genome sequence reveals intriguing examples of molecular mimicry [2]. It was shown that members of the multigene family encoding the KIR proteins have a predicted extracellular domain that shows stretches of identity to host proteins with particularly strong matches to CD99, a human immunoregulatory protein found on the surface of T cells and other lymphocytes. These data raise the interesting possibility that the kir gene products may play a more active role in immune suppression through competition with T cells for CD99 partner molecules rather than just functioning as an antigenic smokescreen, a presumed role for many of the proteins encoded by highly variable Plasmodium multigene families (vars, stevors, virs).

Genetic Regulatory Networks

In organisms that are relatively difficult to genetically manipulate, genomic methods offer opportunities to define regulatory networks by linking motifs in the promoters of co-expressed genes to the DNA-binding activity of different transcription factors. It was recently shown that sets of co-transcribed genes in P. falciparum often share short sequence motifs upstream of their ATGs at rates not expected by chance [5]. A similar approach has been shown to work in Toxoplasma gondii, where functional annotations served as a substitute for gene expression groupings [10]. Although site-directed mutagenesis in P. falciparum has validated the importance of some of these motifs controlling promoter activity, the identity of proteins that bind these motifs has remained generally obscure. However, recently de Silva and coworkers used a protein-binding microarray that contains every possible 10-mer [6] to discover the motifs bound by a series Apicomplexan AP2 transcription factors [7]. These are members of a putative transcription factor family discovered by bioinformatic searches and are homologous to a family in Arabidopsis named the AP2/ERF DNA-binding family [8]. Remarkably, several of these motifs were near perfect matches to the set of motifs shown to be associated with genes involved in invasion or exoerythrocytic stage function in the transcriptional analysis [5]. Moreover, Yuda and coworkers provided genetic confirmation that one of the AP2 proteins regulates genes expressed in the ookinete stage [9] by binding to specific six-base sequences in the proximal promoter. The next challenge will be to perform chromatin immunoprecipitation studies on all DNA-binding proteins and to examine their genome-wide occupancy with a goal of creating a complete map.

Epigenetics of Antigenic Variation

While specific promoter elements are likely to regulate some genes, chromatin structure may play a major role in controlling transcription of genes involved in antigenic variation in multiple parasite species. Malaria parasites and trypanosomes both have large sets of genes that are involved in antigenic variation, and while the two species are well separated on the tree of life, epigenetics appear to control expression of genes involved in antigenic variation in both species. In African trypanosomes, it was shown that a particular histone methylase is responsible in repressing variant surface glycoprotein genes involved in antigenic variation [11]. In malaria parasites, disrupting the histone deacetylase PfSir2A, but not PfSir2B, also results in derepression of genes involved in antigenic variation [12]. Genome-wide chromatin immunoprecipitation studies have also shown correlations between various histone modifications [13],[14], or P. falciparum heterochromatin protein 1 [15] and the location of clonally variant gene families in P. falciparum. Likewise, histone variants mark the start of polycistronic Pol II transcripts in trypanosomes [16]. The patterns of histone occupancy and modification may lead to new theories for how the regulation and switching of antigenic variation genes, critical to pathogenesis, are controlled.

Expression Quantitative Trait Loci

Sexual crosses can be difficult to perform in parasites. Nevertheless, laboratory crosses have been performed on several occasions for T. gondii and P. falciparum. The resulting progeny have been used to map genes involved in drug resistance, host specificity [3], and virulence. While the crosses were usually set up to map a particular trait (e.g., chloroquine resistance), the progeny strains can also be used to map the locus responsible for any quantitative phenotypic difference that separates the two parental lines. Such phenotypes may include growth rate, host cell invasion pattern, differences in the immunolocalization pattern of a given marker, or even gene expression differences [17] that are mapped using a method called expression quantitative trait locus (eQTL) mapping. eQTL work involves the use of linkage mapping to locate genome regions that determine transcript abundance. Both cis loci and trans loci can be identified. An allele that gives rise to a cis eQTL might affect transcript abundance for just that gene by affecting promoter activity or transcript degradation rates, while a trans eQTL, potentially in a transcription factor or an RNA-binding protein, might affect the transcript levels at a variety of unlinked loci. By examining the full genome expression profile of different progeny from a genetic cross, one can determine potential regulatory loci shared by all strains having the same expression phenotype (Figure 1). In P. falciparum, expression studies were performed on a series of progeny clones from a genetic cross between a chloroquine-resistant strain (Dd2) and chloroquine-sensitive strain (HB3) [18]. The authors of this paper identified a powerful trans eQTL on chromosome 5 that controls expression at a large number of genes across the genome and co-localizes with an important drug resistance gene (pfmdr1). However, similar studies using the progeny of a genetic cross between a virulent and less virulent strain of T. gondii only revealed cis-acting loci, indicating that virulence differences were likely to be in polymorphic genes [19] and not in any regulatory factor. Because different host strains are known to be more or less susceptible to parasite infection, the same approach could be used to map regulatory genes controlling the host's response to infection by examining expression profiles of white blood cells or in affected organs in susceptible and nonsusceptible hosts.

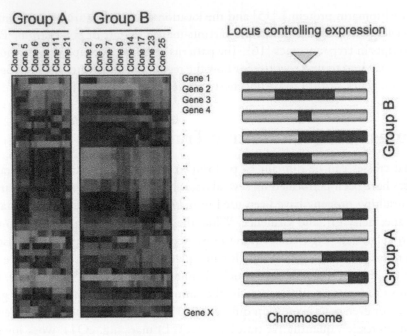

Figure 1. Expression quantitative locus (eQTL) mapping. In this method, different clones from a genetic cross are expression profiled, potentially resulting in two or more different groups, which show distinct expression patterns for a variety of genes as represented by the red-green heat map. Genotyping is then performed on the clones. Loci are identified that are shared by all the clones with the group A pattern, but not by clones with the group B pattern (hypothetical data). The locus may encode a transcriptional regulatory protein that controls the expression of a variety of different genes.

Translating Genomics into Drug Discovery

Over the past several years, the problem of rapidly emerging drug resistance has led to substantial investments in drug discovery programs that have sought to place new drugs for neglected diseases into the pipeline. Drug discovery efforts have benefited from genome sequencing programs that have revealed targets that are found in parasites but are lacking in humans. However, an additional and potentially unrecognized benefit of having parasite genome sequences is that they offer a very powerful approach for rapidly determining an uncharacterized drug's likely mechanism of action or target using in vitro evolution studies. This classic method, which involves growing microbes in sub-lethal concentrations of a drug until they become resistant and then mapping the mutant allele through complementation, has been available to bacteriologists for many years. Because parasites may lack efficient complementation methods, parasitologists have had to wait for the advent of full genome sequencing or the availability of comprehensive full genome tiling arrays to use this approach. Recently, Dharia et al. showed that tiling microarrays, in addition to uses in discovering new transcripts [20] or characterizing variation [21], could be used to detect a copy number variant responsible for fosmidomycin resistance and a newly emerged point mutation responsible for blasticidin resistance [22]. Full genome deep sequencing methods also may give similar results and may be the only option for diploid organisms. Copy number variants or SNPs discovered in the laboratory and associated with drug resistance may eventually be examined in the field. Nair et al. examined linkage disequilibrium with a previously identified copy number variant and showed that GTP cyclohydrolase I amplifications are in linkage disequilibrium with key drug resistance mutations in dihydrofolate reductase [23], suggesting a functional linkage between these two genes.

The frontier of parasite genomics is probably not in sequencing more parasite species or in collecting gene expression data from another pair of conditions. Advances are more likely to be through the integration of large multifaceted datasets, and through studies of complex systems, such as the global transcriptome of the parasite in immune and nonimmune patients, or susceptible and nonsusceptible inbred mice lines. In addition, there are great opportunities for combining population biology with genomics. One could imagine in the future pinpointing the molecular basis of drug resistance through eQTL mapping using expression profiles of parasites obtained from the blood of individuals who had clinically failed treatment. Before this can be realized, however, similar advances in methods for phenotyping parasites will need to be developed. Nevertheless, it seems likely the impact of genomics will soon be measured at the bedside.

References

1. Carlton JM, Adams JH, Silva JC, Bidwell SL, Lorenzi H, et al. (2008) Comparative genomics of the neglected human malaria parasite Plasmodium vivax. Nature 455: 757–763.

2. Pain A, Bohme U, Berry AE, Mungall K, Finn RD, et al. (2008) The genome of the simian and human malaria parasite Plasmodium knowlesi. Nature 455: 799–803.

3. Hayton K, Gaur D, Liu A, Takahashi J, Henschen B, et al. (2008) Erythrocyte binding protein PfRH5 polymorphisms determine species-specific pathways of Plasmodium falciparum invasion. Cell Host Microbe 4: 40–51.

4. Singh B, Kim Sung L, Matusop A, Radhakrishnan A, Shamsul SS, et al. (2004) A large focus of naturally acquired Plasmodium knowlesi infections in human beings. Lancet 363: 1017–1024.

5. Young JA, Johnson JR, Benner C, Yan SF, Chen K, et al. (2008) In silico discovery of transcription regulatory elements in Plasmodium falciparum. BMC Genomics 9: 70.

6. Bulyk ML, Gentalen E, Lockhart DJ, Church GM (1999) Quantifying DNA-protein interactions by double-stranded DNA arrays [see comments]. Nat Biotechnol 17: 573–577.

7. De Silva EK, Gehrke AR, Olszewski K, Leon I, Chahal JS, et al. (2008) Specific DNA-binding by apicomplexan AP2 transcription factors. Proc Natl Acad Sci U S A 105: 8393–8398.

8. Balaji S, Babu MM, Iyer LM, Aravind L (2005) Discovery of the principal specific transcription factors of Apicomplexa and their implication for the evolution of the AP2-integrase DNA binding domains. Nucleic Acids Res 33: 3994–4006.

9. Yuda M, Iwanaga S, Shigenobu S, Mair GR, Janse CJ, et al. (2009) Identification of a transcription factor in the mosquito-invasive stage of malaria parasites. Mol Microbiol 71: 1402–1414.

10. Mullapudi N, Joseph SJ, Kissinger JC (2009) Identification and functional characterization of cis-regulatory elements in the apicomplexan parasite Toxoplasma gondii. Genome Biol 10: R34.

11. Figueiredo LM, Janzen CJ, Cross GA (2008) A histone methyltransferase modulates antigenic variation in African trypanosomes. PLoS Biol 6: e161. doi:10.1371/journal.pbi -o.0060161.

12. Tonkin CJ, Carret CK, Duraisingh MT, Voss TS, Ralph SA, et al. (2009) Sir2 paralogues cooperate to regulate virulence genes and antigenic variation in Plasmodium falciparum. PLoS Biol 7: e84. doi:10.1371/journal.pbio.1000084.

13. Salcedo-Amaya AM, van Driel MA, Alako BT, Trelle MB, van den Elzen AM, et al. (2009) Dynamic histone H3 epigenome marking during the intraerythrocytic cycle of Plasmodium falciparum. Proc Natl Acad Sci USA 106: 9655–9660.

14. Lopez-Rubio JJ, Mancio-Silva L, Scherf A (2009) Genome-wide analysis of heterochromatin associates clonally variant gene regulation with perinuclear repressive centers in malaria parasites. Cell Host Microbe 5: 179–190.

15. Flueck C, Bartfai R, Volz J, Niederwieser I, Salcedo-Amaya AM, et al. (2009) Plasmodium falciparum heterochromatin protein 1 marks genomic loci linked to phenotypic variation of exported virulence factors. PLoS Pathog 5: e1000569. doi:10.1371/journal.ppat.1000569.

16. Siegel TN, Hekstra DR, Kemp LE, Figueiredo LM, Lowell JE, et al. (2009) Four histone variants mark the boundaries of polycistronic transcription units in Trypanosoma brucei. Genes Dev 23: 1063–1076.

17. Gonzales JM, Patel JJ, Ponmee N, Jiang L, Tan A, et al. (2008) Regulatory hotspots in the malaria parasite genome dictate transcriptional variation. PLoS Biol 6: e238. doi:10.1371/journal.pbio.0060238.

18. Schadt EE, Monks SA, Drake TA, Lusis AJ, Che N, et al. (2003) Genetics of gene expression surveyed in maize, mouse and man. Nature 422: 297–302.

19. Boyle JP, Saeij JP, Harada SY, Ajioka JW, Boothroyd JC (2008) Expression quantitative trait locus mapping of toxoplasma genes reveals multiple mechanisms for strain-specific differences in gene expression. Eukaryot Cell 7: 1403–1414.

20. Mourier T, Carret C, Kyes S, Christodoulou Z, Gardner PP, et al. (2008) Genome-wide discovery and verification of novel structured RNAs in Plasmodium falciparum. Genome Res 18: 281–292.

21. Jiang H, Yi M, Mu J, Zhang L, Ivens A, et al. (2008) Detection of genome-wide polymorphisms in the AT-rich Plasmodium falciparum genome using a high-density microarray. BMC Genomics 9: 398.

22. Dharia NV, Sidhu AB, Cassera MB, Westenberger SJ, Bopp SE, et al. (2009) Use of high-density tiling microarrays to globally identify mutations and elucidate mechanisms of drug resistance in Plasmodium falciparum. Genome Biol 10: R21.

23. Nair S, Miller B, Barends M, Jaidee A, Patel J, et al. (2008) Adaptive copy number evolution in malaria parasites. PLoS Genet 4: e1000243. doi:10.1371/journal.pgen.1000-243.

XML-Based Approaches for the Integration of Heterogeneous Bio-Molecular Data

Marco Mesiti, Ernesto Jiménez-Ruiz, Ismael Sanz,
Rafael Berlanga-Llavori, Paolo Perlasca,
Giorgio Valentini and David Manset

ABSTRACT

Background

The today's public database infrastructure spans a very large collection of heterogeneous biological data, opening new opportunities for molecular biology, bio-medical and bioinformatics research, but raising also new problems for their integration and computational processing.

Results

In this paper we survey the most interesting and novel approaches for the representation, integration and management of different kinds of biological data by exploiting XML and the related recommendations and approaches. More-over, we present new and interesting cutting edge approaches for the appropriate management of heterogeneous biological data represented through XML.

Conclusion

XML has succeeded in the integration of heterogeneous biomolecular information, and has established itself as the syntactic glue for biological data sources. Nevertheless, a large variety of XML-based data formats have been proposed, thus resulting in a difficult effective integration of bioinformatics data schemes. The adoption of a few semantic-rich standard formats is urgent to achieve a seamless integration of the current biological resources.

Introduction

Convergent advances in biochemistry techniques, biotechnologies, information technology and computer science provided the basis for the development of bioinformatics and made available huge and growing amounts of biological data [1].

Today's public database infrastructure spans a very large collection of heterogeneous biological data, opening new opportunities for molecular biology, biomedical and bioinformatics research, but raising also new problems for their integration and computational processing. Indeed the integration of multiple data types is one of the main topics in bioinformatics and functional genomics, and several works showed that the integration of heterogeneous bio-molecular data sources can significantly improve the performances of data mining and computational methods for the inference of biological knowledge from the available data [2-5]. In this context a key issue is the representation of the basic bio-molecular entities and biological systems, their associated properties and data in a universal format interchangeable between different databases.

XML [6] has emerged as the most interesting recommendation for the representation and exchange of semi-structured information on the Web. The possibility to easily extend the structure and content of documents as well as the flexible association of schema information makes XML one of the main means for the representation of information exchanged on the Web and, in particular, of biological data. XML also provides a large set of other recommendations, standards and approaches that can be exploited for the representation and management of XML within database systems: query languages (like XPath and XQuery [7]) for querying collections of XML documents and obtaining adequate results; transformation facilities (XSLT [8]), for the presentation of the document contents with different formats (HTML, pdf, doc, etc.); description of schema information

(DTD and XML Schema [9]) to enforce integrity constraints; SQL extension to handle at the same time (object-)relational and XML data (SQL/XML facilities [19]); indexing structures ([20]) for the efficient evaluation of queries. Moreover, many results from both the database and information retrieval communities have been presented for the integration and management of heterogeneous biological data represented through XML. Finally, new general purpose technologies (like Web Services, Grid computing, P2P data management systems) can be exploited to properly process heterogeneous bio-molecular data.

In this paper we first review the principal biological data types that have been identified and analyzed from the biological community and are currently available in different heterogeneous databases. Then, we present different proposals for the XML representation of many biological data types and the main initiatives that exploit XML for the integration of heterogeneous biological data. XML is thus not only employed for the exchange of data on the Web, but also for their management and integration. For what concerns data integration, we point out how conventional and advanced approaches based on Web services and P2P data management systems work specifically on XML and the key points and drawbacks of such approaches. Finally, we envision some future research directions for XML-based heterogeneous bio-molecular data integration, and also emphasize that further knowledge can be integrated with XML in order to overcome its limitations.

Biological Data Types

In this section we introduce the main different types of bio-molecular data and their characteristics, considering also the database infrastructure that houses this information at different levels of representation.

Primary Sequence Data

Historically the first types of data made publicly available have been nucleotide sequence data. It is well-known that EMBL, GenBank and DDBJ host primary sequence data with basic information about the sequence of DNA and RNA [21]. The content of these data bases (DBs) is the same as it constitutes the common base upon which most of the other bio-molecular DBs are built on. This integration effort is due to the international collaboration between the three most important bioinformatics institutions in Europe, USA, and Japan. Nevertheless, problems of accuracy and redundancy of the available entries of these databases can arise. These are due to both the quality of the annotations and biological representation issues (e.g. different Expressed Sequence Tags – EST – sequences are tissue specific and related to the functions of a specific gene). Thus, in some cases it would be necessary to identify such redundancies when dealing with multiple data sources.

Protein DBs represent the second important source of biological sequence data. The SWISSPROT DB is the reference protein bank for the "in silico" analysis of proteins and protein patterns, while TREMBL collects protein sequences obtained by translation from coding nucleotide sequences. Both the primary nucleotide DBs and SWISSPROT store sequence information in flat files, although an XML representation of these files is also available.

Motif and Domain Data

Motifs and protein domains represent bio-molecular entities, usually discovered with pattern recognition methods applied to basic primary sequence data, which are widely used in bioinformatics and molecular biology research to characterize functions and families of proteins. Different specialized databases have been integrated in InterPRO [22], an EBI bioinformatics resource that allows the simultaneous search over different protein domain DBs, through SRS (Sequence Retrieval System) [23] or the Oracle DBMS. Pfam is a DB of families of proteins with common structural and functional elements [24]. They are represented through multiple sequence alignments and Hidden Markov Models. Entries are hierarchically structured from families, to domains, repeats and motifs. Pfam covers also families of proteins obtained through PSI-BLAST [25], an iterative version of the popular BLAST alignment tool for the progressive construction of profiles. The obtained multi-alignments and profiles are stored in the ProDOM DB [26]. Aminoacidic patterns, selected from protein sequences through experimental analysis and computational methods, are available in PROSITE [27]. Each entry of the DB is represented through a description of the pattern, bibliographic links, functional annotation and entries of the SWISSPROT DB where the pattern has been localized. The PRINTS DB represents families of proteins as a hierarchy, where families are related on the basis of their functionalities [28]. Each family is characterized by a "fingerprint", which is a set of conserved motifs deduced from multi-alignments.

Structural Data

Structural data of proteins refer to the atomic spatial coordinates of the atoms and aminoacids composing the protein itself. The reconstruction of the three-dimensional structure of a protein is of paramount importance to understand its function. Data are obtained by X-ray crystallography or NMR spectroscopy. Each entry of the PDB (Protein DataBase) is a file with several records and fields where all the details of the three-dimensional structure of the protein are available, as well as primary and secondary structure information and annotations [29].

Gene Level Data

Although gene databases started with the annotation of primary sequence databases, recent advances in international projects for sequencing entire genomes have promoted the development of specific gene-centric data. For example, Entrez Gene provides a "gene-centered" view of bio-molecular data [30]. For each genetic locus, official gene names and synonyms, together with links to primary DBs are available.

All the information about the context of a specific gene are provided: information about transcripts, products, genomic regions, genotype, phenotype, related pathways and gene ontology terms are linked to the gene under investigation.

KEGG GENES is a collection of gene catalogs for all complete genomes and some partial genomes generated from publicly available resources [31].

This collection is part of KEGG, the Kyoto Encyclopedia of Genes and Genomes and provides a set of integrated databases that can be used to perform system level analyses [32]. KEGG GENES includes the KEGG Orthology (KO) system, a classification system of orthologous genes, including orthologous relationships of paralogous gene groups. Data about orthologous genes coding evolutionarily related proteins in different organisms as well as clusters of paralogous genes conserved in different species are available in COG: these data represent orthologs as clusters of individual proteins delineated by comparing protein sequences encoded in complete genomes [33].

Related DBs are represented by collections of nucleotide patterns with control and regulatory functions. For instance, TRANSFAC is a data bank for transcription factors involved in the regulation and activation of transcription [34]. Data refer to transcription factors and the corresponding DNA binding sites, and can be used for the analysis of gene regulatory events and networks. UTRdb is a database of the untranslated regions of eukaryotic transcripts [35]. They play a fundamental role in post-transcriptional processes of the regulation of gene expression, in the subcellular localization and translation of mRNA. Data related to both the post-translational modification and the regulation of translation are available in TRANSTERM [36].

Genomic Data

The characteristics and properties of bio-molecules can be investigated at the "omics" level: from the study and analysis of single genes or proteins the new biotechnologies introduced at the end of '90s permit to analyze the entire set of genes (genome) or proteins (proteome) of a given species. These data have been generated from the sequencing and mapping of the genome of entire organisms and are available as species-specific resources (e.g. FlyBase for D. melanogaster [37], SGD for S. cerevisiae [38], MGD for M. musculus [39]), or as integrated resources. For instance Ensembl collects data of the human genome and other organisms relative to gene mappings, functional annotations, transcripts, domains, mutations and other relevant information at genomic level [40]. Data are publicly available as flat files. Another similar genomic resource is represented by the Genome Browser [41].

Transcriptomic Data

DNA microarray data collect gene expression levels (i.e. levels of mRNA expressed in a given cell at a given time) at a genome-wide scale [42]. These data allow the analysis of the variability of gene expression between different tissues, individuals, or between different functional or pathological conditions. Three main projects developed at NCBI, EBI and Japan provide access to large collections of these data. GEO, Gene Expression Omnibus, provide structured data for platforms (probes that denote each spot on the array), samples (data of the molecules that need to be analyzed) and series (tables that link samples of an expression experiment to the corresponding platform). GEO is integrated within the NCBI Entrez web site [43]. ArrayExpress, developed at EBI is built on an Oracle DBMS, collects data MIAME-compliant (Minimum Information About a Microarray Experiment) using three main structures: Experiments, Array and Protocols. A subset of curated data can be queried on gene, sample, and experiment attributes [44].

Polymorphism and Mutation Data

Polymorphisms and mutations data are now available in public databases and allow the analysis at genomic level of the associations between mutations and clinic phenotypes [45], as well as studies in the field of population genetics [46]. The database dbSNPs collects data relative to SNPs (Single Nucleotide Polymorphism), region polymorphisms and mutations associated to specific pathologies [47]. Other databases collect bio-medical data for the association between mutations and diseases. For instance HGMD (Human Gene Mutation Database) provides data obtained from literature about mutations and gene alterations related to hereditary diseases, with annotations that associate each mutation to the corresponding clinic phenotype. The OMIM (Online Mendelian Inheritance in Man) database reports data correlated to genetic Mendelian diseases. Data are collected in forms with phenotypes associated to chromosome alterations, to SNPs and mutations, with links to other databases (e.g. Entrez Gene) and cross-references to literature [48]. It is worth mentioning that OMIM provides an XML-based representation to export query results.

System Level Relational Data

The relationships and interactions between different entities and subsystems in cells at different levels (e.g. gene networks or the metabolism of an entire cell) represent a class of relational data by which we can model the behaviour of complex biological systems. These data, mainly obtained through high-throughput bio-technologies, can be used to infer the complex relationships between biomolecules at "system level," considering biological phenomena as the result of the integration of different processes and different interactions involving the entire genome and proteome [49,50].

An example is represented by protein and genetic interaction data collected in BioGRID from major model organism species derived from both high-throughput studies and conventional focused studies [51]. BioGRID houses high-throughput two-hybrid [52] and mass spectrometric protein interaction data [53] and synthetic lethal genetic interactions obtained through synthetic genetic array and molecular barcode methods [54], as well as a vast collection of well-validated physical and genetic interactions from literature. Databases of biological networks offer other examples of relational data that can be used to model regulation processes of gene expression, and post-translational processes related to the metabolism and cellular transport of proteins. For instance the KEGG PATHWAY DB collects different interactions between proteins and genes represented through graphs: e.g. interactions between transcription factor and corresponding target genes, direct interactions (binds) between proteins, or relationship between enzymes participating to the same metabolic process. Other KEGG DBs are obtained by the systematic application of computational biology algorithms to the entire genome of an organism. For instance SSDB is a huge weighted, directed graph, where links corresponds to pairwise comparison of genes using Smith-Waterman similarity scores. The graph can be used to infer orthologs and paralogs or conserved gene clusters or as input to machine learning algorithms to predict gene functions.

Advanced XML-Based Representations of Biological Data

The advent of XML as meta-language able to describe different kinds of data has led to the development of different XML-based languages for the description of biological data types.

In the last few years we have observed the proliferation of XML-based languages for the description of the (1) principal bio-molecular entities (DNA, RNA and proteins) and their structural properties, (2) gene expression (microarray), and (3) system biology. Initial proposals have been developed within small groups of institutes with the main aim of having a common representation of data structures and languages to model their own set of bio-molecular data types, whereas nowadays there are more initiatives (e.g. MIAME) to have a wider general agreement by specifying the minimal requirements that such kinds of data structures and languages should have. Table 1 summarizes some of the characteristics of a subset of existing XML languages (a further discussion on XML standards can be found in [55,56]).

Table 1. XML languages for the representation of biological data types

Type of Data	Format	Concrete Scope	Version	Comments
Molecular entities	BSML [57]	Biological sequences and sequence annotation	v.3.1/2005	Uses DTD. Included in EMBLxml.
	ProXML [58]	Protein sequences, structures and families	v.1.0/2006	Uses XSD. Included within HOBIT formats
	RNAML [59]	RNA sequence, structure and experimental data	v.1.1/2002	Uses XSD
	AGAVE [16]	Biological sequences and sequence annotation	2003	XSD Included in EMBLxml
	Uniprot XSD [121]	Representation of UniProt Records	2004	XSD, Successor of SP (SwissProt) ML format
	EMBLxml [17]	Biological sequences and sequence annotation	v.1.1./2007	Uses XSD. Currently includes BSML and AGAVE.
	GAME [18]	Genome and Sequence	v.0.3/1999	Uses DTD
	SequenceML	Sequence Information	v.2.1 2006	Designed to replace FASTA. Belongs to HOBIT XML formats.
Biological Expression	GeneXML [122]	Gene expression data	-	Uses DTD
	MAGE-ML [123]	Microarray expression data	v.1.0/2006	Uses DTD
System Biology	CellML [124]	Models of biochemical reaction networks	v.1.1/2006	Uses DTD. Available conversion to BioPAX.
	SBML [57]	Models of biochemical reaction networks	Lev. 2/2007	Uses XSD. Available conversion to BioPAX.
	PSI-MI [125]	Protein Interactions	v.2.5/2005	Uses XSD and OBO. Linked with OBO vocabularies.
	BioPAX [60]	Metabolic pathways, molecular interactions	Lev. 3/2008	Uses OWL. Linked to OBO vocabularies.
	CML [126]	Description of Molecules and Reactions	v.2.1./2003	Uses XSD

This table summarizes some of the characteristics of a subset of existing XML languages. In particular, we note the application scope, the number and year of the current version, and comments such as the kind of schema it relies on, or the interaction with other standards.

XML Representation of Bio-Molecular Entities

The Bioinformatic Sequence Markup Language (BSML) [57] describes biological sequences (DNA, RNA, protein sequences) at different granularity levels via sequence data, and sequence annotation. A BSML document usually contains information about how genomes and sequences are encoded, retrieved and displayed. ProXML [58] is used to represent protein sequences, structures and families. A ProXML document consists of an identity section, containing the description of proteins, and a data section, containing properties of such proteins. RNAML [59] has been proposed for the representation and exchange of information about RNA sequences, and their secondary and tertiary structures. A RNAML document can represent RNA molecules as a sequence along with a set of structures that describe the RNA under various conditions or modelling experiments.

XML Representation of Gene Expression

The MAGE project [10] provides a standard for the representation of microarray expression data to facilitate their exchange among different data systems. MAGE mainly consists of: a data exchange model MAGE-OM (Object Model) and a data exchange format MAGE-ML (Markup Language) according to the standardization project groups responsible of the MIAME and MGED Ontology projects.

XML Representations for System Biology

The need to capture the structure and content of bio-molecular and physiological systems lead to develop SBML (the System Biology Markup Language), CellML (the Cell Markup Language), BioPAX (the Biological Pathways Exchange Language) [60] and the set of HUPO-PSI (Proteomics Standards Initiative) formats [55]. SBML is used to encode models consisting of biochemical entities (species) linked by reactions to form biochemical networks, whereas, CellML encodes models consisting of a number of more generic components, each described in their own component elements. BioPAX and HUPO-PSI formats are examples of standards used to represent both structure and semantics of biological data. They are based on the use of ontologies as controlled vocabularies providing a non-ambiguous meaning of the domain.

Integration Initiatives

As showed above, several formats to represent biological data coming from different sources are available. Therefore, as a result, a large collection of heterogeneous biological data is available. This collection claims to be integrated to obtain a comprehensive view of the domain in order to perform analysis and sophisticated queries over the integrated data. cPath [61] has become an interesting initiative to use PSI-MI and BioPAX as standard exchange formats. cPath is an open software for collecting, storing and querying biological pathway data. Biological Databases can be imported and integrated into cPath via PSI-MI and BioPAX. cPath provides a standard web browser frontend and also a XML-based web service API in order to make data available to third-party applications for pathway visualization and analysis.

Biological Data Integration

Biologists usually access different databases through their web interfaces, collect information (usually in text format) they think relevant and finally manually organize them in order to apply their algorithms and thus prove their theories. More and more there is the need to adopt (semi)-automatic approaches for the integration of biological data or rely on framework that help in the data integration process.

The integration of heterogeneous data sources is a traditional database research area whose purpose is to facilitate uniform access to a federation of several data sources. An integrated system provides its users with a global schema in which their views can be defined, along with the mechanisms needed to translate the elements of the global schema into the elements of the corresponding local schema, and vice versa. The heterogeneity of the integrated sources usually causes conflicts that must be resolved by the translation mechanisms in order to produce global results that are correct and complete.

Conflicts can be produced at different levels, namely: physical, syntactic and semantic levels. Currently, the adoption of Internet-based protocols and XML as interchange language has facilitated the integration at physical and syntactic levels. Indeed, XML technology has been formerly aimed at the syntactic integration through the definition of data models (DTD or XSD schemas), query languages (XPath and XQuery) and declarative transformation languages (XSLT). Additionally, recent XML-based formats like RDF and OWL also allow the specification of semantics for the objects to be integrated (ontologies). We remark that XML technology provides the languages for the representation of the information and lacks methods that implement the required integration. Data integration methods, formerly proposed in the database literature, are known as integration architectures. These architectures have been traditionally classified into three main groups: data warehouses, federated and mediated approaches (see Figure 1 for a summary of them).

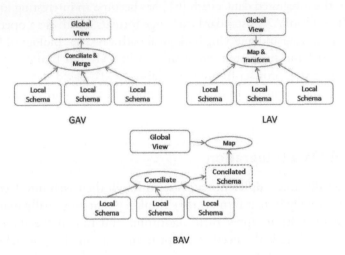

Figure 1. Data integration architectures.

In this section, we will analyse the combination of both XML and data integration architectures for biological data integration. Specifically, we start by introducing the aspects of comparison among the proposed data integration architectures. Then, for each type of architecture, we analyse how proposed systems address such aspects.

Integration Aspects

Table 2 summarizes the main dimensions we regard for comparing current approaches that integrates systems providing biological data. The next paragraphs are devoted to describe them and discuss their relevance.

Table 2. Summary of the integration aspects analyzed in this paper

Aspect	Main approaches
BioData	Sequences, Biological Expressions, Pathways, etc.
Instantiation	Materialized vs. Virtual integration
Integration	Common data storage, data access or data interface
Global View	Local As View, Global As View or Both As View
Global Model	Relational-based, Tree-based, Graph-based
Query Model	Ad-Hoc, SQL, XPath, XQuery, SPARQL, etc.
Semantics	Dictionaries, Thesauri or Domain Ontologies
Scalability	Low (<10 sources), Medium (20–50), High (> 50)

This table represents the aspects around which biological data integration approaches are compared.

BioData

In this aspect we consider the kind of data to be integrated. Some previous papers like [62,63] have analysed the impact of data exchange formats in the integration of biological data and models. All formats rely on XML because of its simple syntax, extensibility and the numerous existing tools for its processing. Among the existing formats, SBML and BioPax are the most accepted ones for integration. As a result, a comprehensive list of converters are available from proprietary formats to SBML/BioPax as well as among themselves.

Instantiation

The degree of instantiation refers to where the physical data reside. In a virtual federation, data reside in the respective data sources, and the integration system gives a unified view of them, whereas in a materialized federation, data are collected from the data sources, cleaned, integrated and stored in a (physically) unique repository. Although the materialized approach is computationally more efficient, in general the virtual approach is chosen because it does not involve data replication, it is more flexible when further data sources should be included in the system, and it is easier to maintain [64].

Integration

The intended degree of integration is also a relevant aspect to take into account when comparing integrated systems. Thus, the integration architecture can be aimed at providing: 1) their common data storage, where biological data are homogenized and consolidated for end users, 2) their common data access, where all users can access (query) homogeneously all the integrated data sources and 3) their common data interface, where users build its tailored integrated applications by combining a series of components that share a common interface (e.g. web services).

Global View

Local As View (LAV) means that the global model has been developed independently from local sources. Afterwards, local data is adapted to the global model in order to give a homogeneous and coherent data representation to end users. Instead, Global As View (GAV) means that the global model has been built by merging local source schemas, unifying entities at two possible levels: schema (S) and instance (I). Hybrid approaches (i.e. Both As View-BAV-) combine both aspects, there is a loosely defined global schema which is mapped to the set of reconciled local schemas (e.g. [65]). Figure 2 illustrates these three ways to generate an integrated global view.

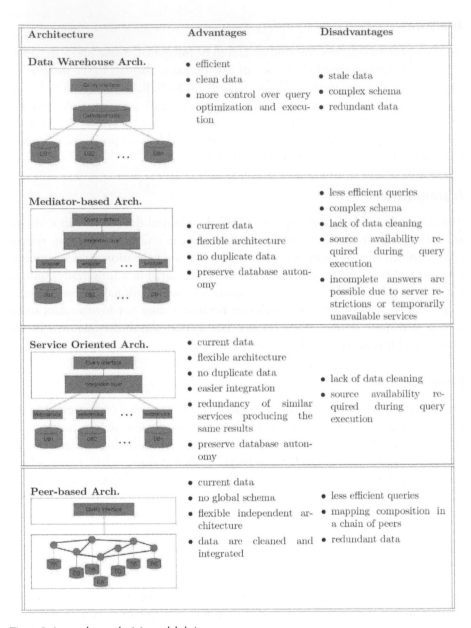

Architecture	Advantages	Disadvantages
Data Warehouse Arch.	• efficient • clean data • more control over query optimization and execution	• stale data • complex schema • redundant data
Mediator-based Arch.	• current data • flexible architecture • no duplicate data • preserve database autonomy	• less efficient queries • complex schema • lack of data cleaning • source availability required during query execution • incomplete answers are possible due to server restrictions or temporarily unavailable services
Service Oriented Arch.	• current data • flexible architecture • no duplicate data • easier integration • redundancy of similar services producing the same results • preserve database autonomy	• lack of data cleaning • source availability required during query execution
Peer-based Arch.	• current data • no global schema • flexible independent architecture • data are cleaned and integrated	• less efficient queries • mapping composition in a chain of peers • redundant data

Figure 2. Approaches to obtaining a global view.

Schema Matching

One of the key issues for building a global view is the generation of mappings between local sources and the global view. In the literature, many approaches for automating the schema matching have been proposed [66]. Basically, a schema matcher is aimed at finding the possible mappings between the elements of two schemas. Such mappings are usually one-to-one but in many cases one-to-many mappings are required. One-to-many mappings are more complex to discover and require some transformation/operation to perform the integration (e.g. current and birth date in a schema must be subtracted to obtain the age in the other schema). Schema matching (SM) has been proposed formerly for relational schemas but it has been also applied to XML and OWL formats. For XML and OWL, SM also regards both the structural constraints and semantic constraints to validate the generated mappings. SM can be used in any of the three approaches: LAV, GAV and BAV. In LAV, SM maps each local source to the global view, in GAV is used to find the unifiable elements of the local sources and in BAV it is used for both.

Regarding Biodata, the use of widely accepted formats like SMBL or BioPax greatly facilitates the generation of global views. SM is partially performed by a manual mapping between SMBL and BioPax (Figure 3). However, a true integration requires a deeper analysis of the values each data record contains. The integration at instance level is also facilitated by the use of external links to well-known resources such as UniProt, OMIM, GeneBank, HUGO, etc. In this case, the integration effort is focused in finding mappings between accession numbers and unique identifiers of these resources [67].

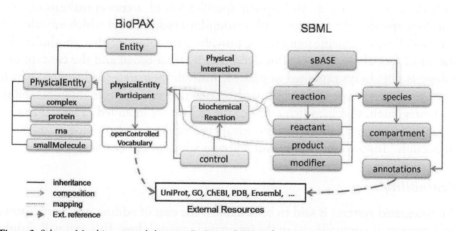

Figure 3. Schema Matching example between BioPax and SBML formats.

Following the schemas in Figure 3, Figure 4 shows examples of possible mappings. In these examples we have used XPath to locate the elements that participate in the mapping. Notice that the first rule involves two entities, the second one two entity attributes and the third one two entities by means of their context (reactants).

$$Species/annotation//.[@resource \sim "uniprot"] \approx Protein$$
$$reaction[@name] \approx biochemicalReaction/Synonyms$$
$$reaction/listOfReactants/.[@species] \approx biochemicalReaction/Left/.$$

Figure 4. Samples of mapping expressions.

Global Model and Query Language

The global model is the representation model for the unified local objects. The more expressive the global model is, the more complex is the global query processing. Traditional approaches rely on relational models (i.e. SQL) which are quite efficient. However, tree-like (e.g. XML) and graph-like (e.g. RDF) models are much more adequate for representing most biological data. The counterpart is that these models present a higher complexity for query processing (e.g. XQuery and SPARQL query processors).

Semantics

Ontologies have been used as mediator schemas defining an abstract layer (semantic level), away from data structures and implementation strategies (physical level), in order to provide a transparent access to heterogeneous resources. Gruber [68] defined ontology as an "explicit specification of a conceptualization." An ontology specifies the concepts and relationships (vocabulary) which are relevant for modelling a domain, moreover it provides a meaning for that vocabulary by means of formal constraints. This definition is rather broad and the concept ontology is not always exploited as desired. Instead, thesauri and glossaries, which have less logical expressivity, are used to facilitate data sources interoperability and integration, that is, which terms of the sources are intended to have the same meaning. Further discussion of the advantages of expressive ontologies are given in Section "Towards more powerful representations of bio-entities."

Scalability

An integrated systems is said to be scalable if the cost of adding new participants (e.g. sources or components) to the integrated system is low. This cost will mainly depend on the difficulty of updating the global view.

Data Warehouse Approaches

A data warehouse integrates and aggregates data of several different DBMSs into a single repository. To this end an integrated database schema is developed that encompasses the schemas of the sources to be integrated. Moreover, views targeted to the analysis to be performed can be realized. Usually an integrated database schema is developed from scratch and can be seldom updated. Updates should be performed sparingly even if, due to a change of user requirements, they are mandatory.

Systems that rely on the data warehouse architecture are usually restricted to consider a few source databases, but can achieve a higher degree of integration of the data sources. The limitation of warehouse system is mainly due to the difficulty to integrate in the system new data sources without changing the schema of the data warehouse. Therefore, these systems allow to obtain an high degree of instantiation. Examples of these systems are the following ones:

- DWARF [69], which integrates data on sequence, structure, and functional annotation for protein fold families. DWARF extracts data from public available resouces (e.g. GenBank, ExPDB and DSSP).

- BioWarehouse [70] is an open source toolkit for constructing bioinformatics database warehouses by integrating a set of different biological databases into a single physical DBMS (MySQL or Oracle). It supports data related to the following types of biological objects: genes and genomes, proteins, enzymatic reactions, biological pathways, taxonomies, nomenclatures, microarray gene expression, computationally-generated results.

- Atlas [71] locally stores and integrates biological sequences, molecular interactions, homology information, functional annotations of genes, and biological ontologies.

- Biozone [72,73] is a unified biological resource on DNA sequences, proteins, complexes and cellular pathways. Biozon combines graph model and hierarchical class approaches to express and characterize biological entities in terms of constraints depending on the relations with other modelled entities or depending on the proper nature of each individual entity. Biozon supports derived data strategies based on similarity relationships and functional predictions enabling propagation of knowledge and allowing the specification of complex queries.

- cPath [61] is an open source database software for collecting, storing and querying biological pathway data. Multiple databases can be imported and integrated into cPath via PSI-MI and BioPAX standard exchange formats. cPath data can be viewed by means of a standard web browser or exported via an XML-based web service API, making cPath data available to third-party applications for pathway visualization and analysis.

Most of these approaches take the LAV strategy to build the global view, and provide a common data and storage model (see Table 3). Due to the complexity of the data loaders, where transformations between schemas are usually hard coded (e.g. Java, C++ and Perl programs), the cost of adding new sources is high. This problem can be alleviated if data sources already provide their data in standard XML formats, in which case a few data loaders (e.g. a BioPax data loader) can deal with many sources. However, any evolution in either the exchange formats or the source schemas will imply a re-implementation of all these loaders, so the cost of maintaining these integrated systems can be very high.

Table 3. Data warehouse approaches

Aspect	DWARF	BioWareh.	Atlas	Biozone	CPath
BioData	Sequences	All Types	Genes	All Types	AllTypes
Instantiation			Materialized		
Integration			Common Storage/Access		
Global View		LAV		GAV (I)	LAV
Global Model		Relational		Graph	RDF/OWL
Query Model		SQL		SQL/AdHoc	SPARQL
Semantics	-	Thesaurus	-	-	Ontologies
Scalability	Low	Medium	Medium	Medium	Medium

This table compares the datawarehouse approaches relying on the aspects introduced in Table 2.

Mediation Approaches

In contrast with data warehouse-based architectures, in mediator-based systems (originally proposed by Gio Wiederhold [74]) individual data sources maintain their independence. Data integration is achieved by defining a global view, or integrated schema, which is shared by all sources; a "mediator" component, or mediator-based middleware, adapts queries formulated against the global view to the local data and capabilities. Typically, each individual source will also require the definition of a "wrapper" component, which will be used to export a view of the local data in a useful format for mediation (by translating the data to/from XML, for instance). Figure 1 depicts a typical mediator-based architecture. Query processing is achieved by sending subqueries to relevant sources, and then combining the local query results.

Thus, the main advantages of mediator-based architectures are threefold: (i) the insurance that returned data are always up-to-date, since queries are performed dynamically (ii) data are not duplicated since they reside in their local repository, (iii) it is easier to add new sources of information. A major drawback of mediator-based system is the need to manually specify the mappings between local and global schemas; several techniques have been proposed to automate these steps (e.g. [66]).

The following systems are examples of mediator-based systems for biological data:

- Ontofusion [75] proposes a multiple ontology approach to integrate genomic and clinical databases at the semantic level. For each data source an ontology (named virtual schema) is created to describe the structure of the data. Virtual schemas are unified (i.e. merged) in a unique global schema to give an homogeneous access to data.

- TAMBIS [76], unlike Ontofusion, adopts a unique ontology approach to provide a common access to several data resources so that cross database searches seem to be transparent. An ontology called TAO (Tambis Ontology) has been created for this purpose. TAO collects all the requirements of the database to be integrated. Scalability, when adding new resources, is the major drawback of this approach.

- BioMediator [77] uses a logic-oriented knowledge base to store meta-information about each data source, which allows the specification of tailored mediated schemas including rich relationships. The mediator component is extensible through the use of plug-ins, which allows the definition of mapping rules for the tailored schema.

Table 4 summarizes the main mediator-based approaches. Last two columns report the characteristics of two recent internet-based architectures that facilitates the integration of systems: Web Services and Peer-to-Peer architectures. Both architectures are discussed in the next sections.

Table 4. Mediator-based Approaches

Aspect	Ontofusion	TAMBIS	Biomed.	WS	P2P
BioData Instantiation	Genes	All types	Genes	All Types	All Types
Integration			Virtual Common Access		
Global View	GAV (S/I)	GAV (S)	LAV	LAV	N.A.
Global Model	RDF/OWL		XML	RDF/OWL	XML
Query Model	Boolean	CPL	XQuery	SPARQL	XQuery
Semantics	Ontologies		-	-	-
Scalability	Medium	Low	Medium	High	High

This table compares the mediator-based approaches relying on the aspects introduced in Table 2.
(Biomed. = Biomediator, WS = Web Services approaches, P2P = peer-to-peer approaches).

In general, in the Bioinformatics area, mediator-based approaches are less popular than data warehouse ones. One possible reason for this is that mediator-approaches require reversible transformations in order to both distribute global queries to local sources and translate local results as global objects. Data warehouse approaches only require unidirectional transformations (i.e. from local to global view), which makes their implementation easier.

Service-Oriented Architectures (SOAs)

In the previous sections we have been mainly concerned with the integration of biological data sources through the classical data warehouse and mediator approaches. However, Bioinformatics research usually implies processing all these data by means of software applications as those that realize in silico experiments. In this context, Service-Oriented Architecture (SOA) provides a standard method to integrate both data sources and software applications by regarding them as interoperable services. Thus, client applications will combine these services to implement their intended tasks. In this section, we review the main efforts in providing such services within the Bioinformatics community.

Figure 5 shows an abstract Web Service (WS) for retrieving pathways given a set of possible participants. It is represented with a box with three parts: the input, the method name and the output or result. This web service can take part of either a mediator-based architecture (top right part of the figure) or a workflow (bottom part of the figure). However, in order to use concrete web services (i.e. web services located at some machines with a specific interface), applications and users must be aware of the XML schema of input and output parameters. This schema is expressed with the Web Service Description Language (WSDL). Thus, the main integration issue consists of reconciling the schemas of the services to be combined. Biological research institutions like the National Center for Biotechnology Information (NCBI) and the European Institute for Bioinformatics (EBI) have published most of their applications and data sources as Web Services. Thus, researchers can freely invoke the Entrez e-utilities, the EMBOSS suite [11], the EMBL-EBI tools [12] and Distributed Annotation System [13] among others. These Web Services constitute the basic layer over which more complex services and workflows can be defined.

Semantic Web Services

WSDL files have found very limited usage for processing and distributing biological data. As a consequence, new protocols have been proposed to extend the basic functionalities of bioinformatics Web Services. BioMOBY [78] has been quite successful as such an extension. MOBY services are registered in a central node by properly annotating their interfaces. Such annotations mainly involve the input and output data of each service as well as some descriptions about its functionalities. Currently, there are more than 1000 services registered and more than 500 data types associated to their descriptions (see http://sswap.info). Notice that the ratio between data types and services indicates that a further data integration effort should be done in order to make them more interoperable.

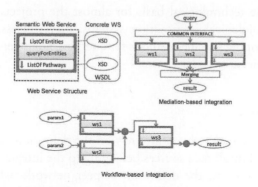

Figure 5. Integration through Web Services.

Workflows

Several proposals have recently appeared to define complex workflows over Bio-MOBY services to perform for example in silico experiments. The most popular of these proposals is the Taverna tool [79], which has been proposed within the myGRID project [80]. This tool allows users to first define graphically a workflow (i.e. chain of service invocations) and then execute it over a GRID-based middleware. Other similar Web-based tools have been proposed, for example MOWServ [81], SeaHawk [82] and Remora [83] to mention a few. Recently, some extensions to the BioMOBY protocol have been proposed according to the new requirements arisen from workflow management [84].

Grid-Based Services

Grid technologies are intended to provide highly scalable computing frameworks where resource-hungry applications can be performed efficiently. As the biological community is continuously generating vast amounts of biological data, which also require time-consuming processes to be analyzed, Grid computing has been usually taken up in large bioinformatic projects (e.g. myGRID, caBIG, EGEE, etc.) Grid technologies also rely on Service-Oriented Architectures. Indeed, recent standards for Grid architectures basically extend the Web Service technology. Thus, the Web Service Resource Framework (WSRF) is the WS extension proposed for the Open Grid Service Architecture (OGSA). Unlike Web Services, Grid services must account for security, transaction and distribution issues arisen from Grid architectures. A good review of Grid technologies applied to Bioinformatics can be found in [85].

Service-Oriented Architectures have an increasingly prominent role in the development of biologiocal data processing and integration. As a result, SOAs are constituting the technological basis for almost the projects aimed at seamlessly integrating biological information systems. Nevertheless, little work has been done in developing specific methods for querying homogeneously biological data-providers services.

Peer Architecture

All the previous presented architectures rely on the definition of a global schema that is well accepted by all data sources belonging to the integrated systems. Current efforts are devoted to the definition of peer networks where data can be locally organized and managed [86]. Each peer or group of peers can share the same schema, and local mapping among pairs of schemas can be established leading to the formulation of a semantic network. When a new peer wishes to join the semantic network, it should establish a mapping simply with a single peer or a subset of the network peers. When a query is submitted to one of the network peers, the query is routed to the peers that, according to the resource availability policies, can contain possible answers. Relying on the pre-established mappings among schemas, it is possible to translate a query to be executed in local schema (and thus obtain more precise results) or to translate the results in order to make homogeneous and comparable the different results. The peer, that initially received the query, is in charge of collecting the answers and returning them to the requesting user or application. Key features of a peer architecture is, thus, the lack of a global huge schema. Peers can develop schemas that are tailored for their main users and then establish a mapping with a small fraction of other peers. A peer can easily join and leave the network. The main drawback of this architecture is the need to develop mappings and their use on the fly to evaluate queries that can effect the performance of the retrieval process. Many efforts are currently devoted to quickly perform these tasks (e.g. developing mapping tables [87]). As in the other architecture, XML plays a central role in semantic peer networks, XML can be exploited both as a message exchange format among peers as well as a format for the representation of the peer contents.

Well-known and general purpose P2P data management systems (PDMS) like Hyperion [86], PeerDB [88], and GridVine [89] have been proposed that rely on the relational model and can be exploited for the management of biological data that do not present complex structures. Moreover, the Bioscout system [90] has been developed for helping biologists in the graphical specification of queries and for developing efficient query plans to be executed in a peer network. Apart from these few systems, P2P technology has been scarcely applied to the biological research.

From a practical point of view, there are not big differences between Service-Oriented Architectures (SOA) and P2P. Both have as strongest point their good scalability. However, unlike SOAs, P2P systems lack a solid and standard technological background (e.g. SOAP, WSLD, OGSA etc.) that makes them fully interoperable.

Advanced issues in XML-based biological data integration

Even if several XML-based approaches for the integration of bio-molecular data have been proposed, several items remain open for current and future research. For instance, XML is mainly employed for the exchange format and in many cases the data management facilities (XSLT, XQuery, indexing structure,...) are not yet exploited. Besides this basic limitation, there are some other important issues in data integration which are not addressed by these systems:

- Data security and privacy. Data contains sensitive information about people that needs to be protected from unauthorized users. Specific approaches are required for biological data because they contain personal characteristics that can lead to the identification of a subject and their obfuscation can alter the experimental results.

- Evolution of data. Biological databases quickly change [91]: data formats, access methods and query interfaces are not stable over time, and even when elaborate database integration solutions are used, a significant amount of time is spent to address this issue.

- Efficiency. Approaches for the efficient evaluation of queries in a distributed and heterogeneous environment as well as approaches for collecting and normalize answers produced from independent sources should be developed.

- Approximation. The richness of data format and organization requires the development of systems that return approximate answers to an user query.

We have to remark that conflicts at physical and syntactic levels are almost solved exploiting XML technologies. However conflicts at the semantic layer are still an open issue for seamless biological data integration.

In the remainder of the section we present the main research initiatives that are currently devised to face these issues in the XML context.

Towards More Powerful Representations of Bio-Entities

Despite the current standardization efforts, the Bioinformatics community still lacks of a standard exchange language and vocabulary for all the biological data.

As shown along this paper, XML-like representations have been widely accepted to represent biological data. Additionally, several controlled vocabularies (e.g. thesauri) are now available to properly annotate these data. These vocabularies are usually expressed in the Open Biological Ontologies (OBO) [14], for example the Gene Ontology, the NCBI Taxonomy, the Cell Ontology, etc. The main drawbacks of these standards are that pure XML representations do not account for semantics, and that OBO ontologies are in most cases limited to simple taxonomies (i.e. informal *is-a* relationships).

The use of more expressive logics would give rise to more powerful and extensible ontologies so that biological concepts can be described not only with taxonomical relationships but also with logical descriptions (axioms). Consider, for example, the following pair of axioms

$$\exists\, participant. \top \sqsubseteq Interaction \tag{1}$$

$$GeneticInteraction \sqsubseteq\, \geq 2\, participant.Gene. \tag{2}$$

It can be derived that Genetic Interaction $\boxtimes$ Interaction, that is, a new implicit is-a relationship is inferred from concept definitions. Notice that in this way, ontologies can be more compact and legible as concept descriptions are closer to natural language expressions.

To the best of our knowledge, BioPax is the only standard relying on an expressive ontology language. BioPax describes biological pathways and their components in the Ontology Web Language (OWL) [15]. In this way, specific pathway data can be classified according to the BioPax concepts by using a reasoner, as long as these data are represented as OWL individuals. It is worth mentioning that in OWL individuals do not need to be explicitly associated to a specific concept, but just to a proper description. This allows biologist to delegate the final classification to a reasoner. For example, taking into account the axioms 3 to 6 involving a set of individuals and the axioms 1 and 2, a reasoner is able to infer that interaction_1 is an individual of the concept GeneticInteraction.

$$participant(interaction\_1,BNI1) \tag{3}$$

$$participant(interaction\_1,ATS1) \tag{4}$$

$$BNI1:Gene \tag{5}$$

$$ATS1:Gene. \tag{6}$$

Ontology-based data integration has been tested in systems like Ontofusion and Tambis previously presented. Ontofusion adopts a multiple ontology approach (e.g. one per source) whereas Tambis uses a unique global schema. Multiple ontology approaches are more scalable since they do not require a global ontology dependent of the data sources. However the implementation and integration is harder since the ontologies of each source should be integrated, that is, mappings between them have to be defined. This task may be rather difficult [92] if ontologies use different names or naming conventions to refer to their entities. Assuming that ontologies can be easily mapped (e.g. they use common vocabulary) semantic compatibility still arises as an open issue in ontology integration approaches. Ontologies to be integrated, and therefore the data sources, may contain conflicting descriptions which should be detected to perform a proper integration. This apparently disadvantage of the multiple ontology approach could also be seen as a strong, since ontologies could be exploited to detect those incompatibilities between data sources and then to repair/adapt them to make possible the integration. When integrating ontologies errors and incompatibilities manifest themselves as unintended logical consequences (e.g. unsatisfiable concepts or unintended subsumptions). In the literature several approaches can be found to detect and repair unintended logic consequences [93-95]. These techniques localize those sets of descriptions (i.e. axioms) which provoke the error (i.e. incompatibility).

Nevertheless, although the use of expressive ontologies seems to be a feasible solution to both the semantic representation of data sources and the classification of biological data, in practice, they are not being adopted as expected. The design of expressive ontologies requires strong skills in Description Logics (DL) [96], which are not familiar to biologists. That is why less expressive languages like OBO has become so popular among biologists.

Open Issues in Service Oriented Architectures

The use of Web Services in Bioinformatics have been earlier analyzed in [97]. Some of the issues reported in this paper are being currently addressed, for example: the migration of HTML-based query forms to web service interfaces, the improvement of the discovery tools for biological web services (e.g. Semantic Bio-MOBY), and the overhead produced by XML when dealing with large biological data objects. However, there are some other issues that still remain open. Among them, we emphasize those related to data integration, namely:

- Web service architectures allow biologists to have several alternative sources for the information they request. In contrast, the selection of the proper sources will depend on criteria that are not usually found in these architectures, like the authority of the provider, the version of the data collection behind the service, etc. In this way, new metadata should be defined to guide users in the selection of the services they require for their tasks.

- Workflows also require some criteria and methods to select the services that potentially can comprise them [98]. These criteria must go beyond simple annotations of input/output parameters, because compositions can require more complex interactions between the involved services. For example, non-trivial data transformations may be required in order to connect two web services (i.e. Mediators). Additionally, we need the discovery of semantic mappings between WS data types to look for further potentially compatible services.

- Biological web services require an integrated data space consisting of just a few standard data formats, instead of the hundreds XML data types currently available. In this way, any data type used in a web service should be defined within a widely accepted semantic-based standard (e.g BioPax).

Approximate Retrieval of Information

As earlier commented, data warehouse approaches allow a high degree of integration but at the cost of complying with a common database schema, which makes it difficult the inclusion of new data sources or the evolution of existing ones. Recently, several research works proposed to create XML data warehouses with data published in the Web (see [99] for a review). Basically, XML warehouses propose to store the XML data as it is without imposing any common schema. Afterwards, by applying clustering techniques and XML schema inference methods, the data warehouse provides the proper structures to support data exploration and analysis. However, these systems should face the high heterogeneity the stored XML data may present. Unfortunately, well-known XML tools like XPath and XQuery are not appropriate in this context, because they assume a well-defined schema.

Current approaches to handle highly heterogeneous XML collections are based on both approximate query processing [100,101] and multi-similarity systems [102]. The former consists of defining a relaxed query (pattern) in order to retrieve a list of similar XML documents (fragments). The latter ones provide multiple notions of similarity simultaneously in order to account for the heterogeneity of the data contained in the stored XML documents. The ArHex system [101] combines both methods in order to provide an extensible framework where users can adjust their similarity measures to the collection complexity. Such a framework could be used as the basis for defining novel exploration and analysis tools over highly heterogeneous biological data sets.

Evolution of Data

The rapid development of technologies leads to quickly change both biological data and applications working with such data.

For what concerns data, different problems should be faced. The introduction of new versions of data structures already developed leads to the problem of their management and also to determine the version on which queries should be evaluated. The evolution of data structures may imply the elimination of the old versions of data, but it introduces the issue of modifying existing instances in order to adhere to the evolved structures.

For what concern applications, the evolution of data structures requires to update the applications working on them in order to work properly with the different versions as well as the evolved structures. Moreover, mapping among schemas of two sources, when one of the two is modified, needs to be adapted.

The representation of biological data in the XML format can introduce further issues when modifying the schema (either represented through a DTD or a XML Schema). Specifically, the evolution of a schema may lead to revalidate documents already developed according to the old schema to check whether they are still valid for the new schema and, whenever they are no longer valid, to adapt the documents to the new schema. In [103], the X-evolution framework has been presented to address the issue of XML schema evolution. The authors propose both graphical and query-based approaches for the specification of schema modification and for adapting the documents to the new schema. Nevertheless, more specific approaches adapted to biological data should be addressed.

Schema modifications also impact on applications, queries, and mappings between schemas. The impact of schema evolution on queries and mappings has been investigated ([104-106]). The issue of automatically extending applications working on the original schema when this has evolved has not been addressed in the context of XML.

Last, but not least, another issue to be faced is ontology evolution; that is, the issue of modifying an ontology in response to a certain change in the domain or its conceptualization. The issues of ontology mapping, alignment, and evolution and their consequences on ontology instances should be addressed in the highly evolving context of biological data ([107-109]).

Security and Data Privacy

The integration and management of heterogeneous data sources into a huge and organized data repository supports the scientists in making and proving the validity of their theories but it also produces as a side-effect the opportunity for a malicious user to access to or to make a prediction about relevant sensitive data. As an example, in healthcare domain a malicious user may be interested in patient genomic information in order to predict its current and future health status.

The degree of relevance of data and the kind of countermeasures to adopt in order to react against a malicious attack depend on several different aspects mainly based on the characteristics of the context to be considered and on the type of the attack.

Several approaches have been recently proposed to increase privacy and security in different context [110–114]. Access control, authentication, policy specification and enforcing techniques [115–117] are used to filter the requests to the sensitive resources so that the access requests coming from unauthorized parties be discarded and data be accessible only by users according to the enforced security policy. On the other hand, data obfuscation and data hiding techniques [118–120] are used to preserve privacy and security in presence of data mining techniques and they are based on the idea to distort or encrypt confidential data so that relevant information can not be easily retrieved.

When the security level increases, by adopting different security techniques coexisting together, the data sharing level decreases. Indeed, data are not publicly available but accessible only by those holding the required security credentials. A right tuning of these levels is desirable in order to satisfy both the security requirements and the data sharing demand.

Conclusion

In this paper we pointed out the main current technologies that can be exploited for the integration and management of biological data through XML. We outlined the proposals for the representation of biological data in XML and discussed new interesting approaches that have been emerging in the last few years. We can conclude that XML has succeeded as the syntactic glue for biological data sources. Nevertheless, XML-based approaches produced a great variety of data formats, which makes it difficult to effectively integrate them. The adoption of a few semantic-rich standard formats is urgent to achieve a seamlessly integration of the current biological resources.

Competing Interests

The authors declare that they have no competing interests.

Authors' Contributions

The group of work is formed by Italian and Spanish teams. The idea and structure of this work has been proposed by MM. The Italian team was coordinated by MM, whereas the Spanish team was coordinated by EJR and RB. GV took care of the biological data types. MM, EJR, and PP presented the approaches for the XML representation of biological data. RB, EJR, IS, MM, PP, and DM worked on the data integration issues and approaches within the biological context. All authors jointly proposed the future work, read, and approved the final manuscript.

Acknowledgements

This work has been partially funded by the Spanish National Research Program (contract number TIN2008-01825/TIN). Ernesto Jimenez-Ruiz was supported by the PhD Fellowship Program of the Generalitat Valenciana.

References

1. Galperin M: The Molecular Biology Database Collection: 2008 update. Nucleic Acids Res 2008, 36(Database issue):D2–D4.

2. Pavlidis P, Weston J, Cai J, Noble W: Learning gene functional classification from multiple data. J Comput Biol 2002, 9:401–411.

3. Troyanskaya O, et al.: A Bayesian framework for combining heterogeneous data sources for gene function prediction (in Saccharomices cerevisiae). Proc Natl Acad Sci USA 2003, 100:8348–8353.

4. Lanckriet G, et al.: A statistical framework for genomic data fusion. Bioinformatics 2004, 20:2626–2635.

5. Barutcuoglu Z, Schapire R, Troyanskaya O: Hierarchical multi-label prediction of gene function. Bioinformatics 2006, 22(7):830–836.

6. W3C: Extensible Markup Language (XML) 1.0 (Fourth edition). 2006.

7. W3C: XQuery 1.0 and XPath 2.0 Data Model (XDM). 2007.

8. W3C: XSL Transformations (XSLT). 1999.

9. W3C: XML Schema. 2000.

10. MAGE project [http://www.mged.org/Workgroups/MAGE/mage.html].

11. EMBOSS suite [http://emboss.sourceforge.net/].

12. EMBL-EBI tools [http://www.ebi.ac.uk/services/].

13. Distributed Annotation System [http://www.biodas.org].

14. Open Biological Ontologies [http://www.obofoundry.org/].

15. W3C Ontology Web Language [http://www.w3.org/TR/owl-guide].

16. AGAVE [http://www.agavexml.org/].

17. EMBLxml [http://www.ebi.ac.uk/embl/xml/].

18. GAME [http://xml.coverpages.org/game.html].

19. Melton J, Buxton S: Querying XML – XQuery, XPath, and SQL/XML in Context. Morgan Kaufmann; 2006.

20. Catania B, Maddalena A, Vakali A: XML Document Indexes: A Classification. IEEE Internet Computing 2005, 9(5):64–71.

21. Kulikova T, et al.: EMBL Nucleotide Sequence Database in 2006. Nucleic Acid Res 2007, 35:D16–D20.

22. Mulder N, et al.: New developments in the InterPro database. Nucleic Acids Research 2007, 35:D224–228.

23. Zdobnov E, Lopez R, Apweiler R, Etzold T: The EBI SRS server-recent developments. Bioinformatics 2002, 18(2):368–73.

24. Finn R, Tate J, Mistry J, Coggill P, Sammut J, Hotz H, Ceric G, Forslund K, Eddy S, Sonnhammer E, Bateman A: The Pfam protein families database. Nucleic Acids Research 2008, 36:D281–D288.

25. Altschul S, Madden T, Schaffer A, Zhang J, Zhang Z, Miller W, Lipman D: Gapped Blast and PSI-Blast: a new generation of protein database search programs. Nucleic Acids Research 1997, 25(17):3389–3402.

26. Corpet F, Servant F, Gouzy J, Kahn D: ProDom and ProDom-CG: tools for protein domain analysis and whole genome comparisons. Nucleic Acids Research 2000, 28:267–269.

27. Hulo N, Bairoch A, Bulliard V, Cerutti L, Cuche B, De Castro E, Lachaize C, Langendijk-Genevaux P, Sigrist C: The 20 years of PROSITE. Nucleic Acids Research 2008, 36:D245–D249.

28. Attwood T: The PRINTS database: a resource for identification of protein families. Brief Bioinform 2002, 3(3):252–263.

29. Berman H, Henrick K, Nakamura H, Markley J: The worldwide Protein Data Bank (wwPDB): ensuring a single, uniform archive of PDB data. Nucleic Acids Research 2007, 35:D301–303.

30. Maglott D, Ostell J, Pruitt K, Tatusova T: Entrez Gene: gene-centered information at NCBI. Nucleic Acids Research 2005, 33:D54–D58.

31. Kanehisa M, Goto S: KEGG: Kyoto Encyclopedia of Genes and Genomes. Nucleic Acids Research 2000, 28:27–30.

32. Kanehisa M, Araki M, Goto S, Hattori M, Hirakawa M, Itoh M, Katayama T, Kawashima S, Okuda S, Tokimatsu T, Yamanishi Y: KEGG for linking genomes to life and the environment. Nucleic Acids Research 2008, 36:D480–D484.

33. Tatusov R, Fedorova N, JD J, Jacobs A, Kiryutin B, Koonin E, Krylov D, Mazumder R, Mekhedov S, Nikolskaya A, Rao B, Smirnov S, Sverdlov A, Vasudevan S, Wolf Y, Yin J, Natale D: The COG database: an updated version includes eukaryotes. BMC Bioinformatics 2003, 4.

34. Wingender E: TRANSFAC project as an example of framework technology that supports the analysis of genomic regulation. Brief Bioinformatics 2008, 9(3):326–332.

35. Mignone F, Grillo G, Licciulli F, Iacono M, Liuni S, Kersey P, Duarte J, Saccone C, Pesole G: UTRdb and UTRsite: a collection of sequences and regulatory motifs of the untranslated regions of eukaryotic mRNAs. Nucleic Acids Research 2005, 33:D141–D146.

36. Dalphin M, Brown C, Stockwell P, Tate W: The translational signal database, TransTerm, is now a relational database. Nucleic Acids Research 1998, 26:335–337.

37. Wilson R, Goodman J, Strelets V, the FlyBase Consortium: FlyBase: integration and improvements to query tools. Nucleic Acids Research 2008, 36:D588–D593.

38. Fisk D, Ball C, Dolinski K, Engel S, Hong E, Issel-Tarver L, Schwartz K, Sethuraman A, Botstein D, Cherry J: Saccharomyces cerevisiae S288C genome annotation: a working hypothesis. Yeast 2006, 23(12):857–865.

39. Bult C, Eppig J, Kadin J, Richardson J, Blake J, the members of the Mouse Genome Database Group: The Mouse Genome Database (MGD): mouse biology and model systems. Nucleic Acids Research 2008, 36:D724–D728.

40. Birney E, et al.: An Overview of Ensembl. Genome Res 2004, 14(5):925–928.

41. Karolchik D, et al.: The UCSC Genome Browser Database: 2008 update. Nucleic Acids Res 2008, 36(Database issue):D773–D779.

42. Eisen M, Spellman P, Brown P, Botstein D: Cluster analysis and display of genome-wide expression patterns. PNAS 1998, 95(25):14863–14868.

43. Barrett T, Troup D, Wilhite S, Ledoux P, Rudnev D, Evangelista C, Kim I, Soboleva A, Tomashevsky M, Edgar R: NCBI GEO: mining tens of millions of expression profiles. Nucleic Acids Research 2007, 35:D760–D765.

44. Parkinson H, Kapushesky M, Shojatalab M, Abeygunawardena N, Coulson R, Farne A, Holloway E, Kolesnikov N, Lilja P, Lukk M, Mani R, Rayner T, Sharma A, William E, Sarkans U, Brazma A: ArrayExpress, a public database of microarray experiments and gene expression profiles. Nucleic Acids Research 2007, 35:D747–D750.

45. Giardine B, Riemer C, Hefferon T, Thomas D, Hsu F, Zielenski J, Sang Y, El-nitski L, Cutting G, Trumbower H, Kern A, Kuhn R, Patrinos G, Hughes J, Higgs D, Chui D, Scriver C, Phommarinh M, Patnaik S, Blumenfeld O, Gottlieb B, Vihinen M, Valiaho J, Kent J, Miller W, Hardison R: PhenCode: connecting ENCODE data with mutations and phenotype. Hum Mutat 2007, 28(6):554–562.

46. Huerta-Sanchez E, Durrett R, Bustamante CD: Population Genetics of Poly-morphism and Divergence Under Fluctuating Selection. Genetics 2008, 178:325–337.

47. Sherry S, Ward M, Kholodov M, Baker J, Phan L, Smigielski E, Sirotkin K: dbSNP: the NCBI database of genetic variation. Nucleic Acids Research 2001, 29:308–311.

48. McKusick V: Mendelian Inheritance in Man and its online version, OMIM. Am J Hum Genet 2007, 80(4):588–604.

49. Dalphin M, Brown C, Stockwell P, Tate W: All systems go. Nature 2007, 446:493–494.

50. Kaneko K: Life: An Introduction to Complex Systems Biology. Berlin: Springer; 2006.

51. Stark C, Breitkreutz B, Reguly T, Boucher L, Breitkreutz A, Tyers M: BioGRID: a general repository for interaction datasets. Nucleic Acids Res 2006, 34:D535–D539.

52. Uetz P, Giot L, Cagney G, Mansfield T, Judson R, Knight J, Lockshon D, Narayan V, Srinivasan M, Pochart P, et al.: A comprehensive analysis of protein-protein interactions in Saccharomyces cerevisiae. Nature 2000, 403:623–627.

53. Ho Y, Gruhler A, Heilbut A, Bader G, Moore L, Adams S, Millar A, Taylor P, Bennett K, Boutilier K, et al.: Systematic identification of protein complexes in Saccharomyces cerevisiae by mass spectrometry. Nature 2002, 415:180–183.

54. Davierwala A, Haynes J, Li Z, Brost R, Robinson M, Yu L, Mnaimneh S, Ding H, Zhu H, Chen Y, et al.: The synthetic genetic interaction spectrum of essen-tial genes. Nature Genet 2005, 37:1147–1152.

55. Harvey S, et al.: Standards for systems biology. Nat Rev Genet 2006, 7:593–605.

56. Strömbäck L, et al.: A review of standards for data exchange within systems biol-ogy. Proteomics 2007, 7:857–867.

57. Hucka M, et al.: The Systems Biology Markup Language (SBML): A Medium for Representation and Exchange of Biochemical Network Models. Bioinfor-matics 2003, 19(4):524–531.

58. Hanisch D, et al.: ProML – the Protein Markup Language for specification of protein sequences, structures and families. Silico Biology 2002, 2(3):313–324.

59. Harvey S, et al.: RNAML. A standard syntax for exchanging RNA information. Silico Biology 2002, 8(6):707–717.

60. Bader GD, Cary MP: BioPAX – Biological Pathways Exchange Language Level 2, Version 1.0. [Http://www.biopax.org/release/biopax-level2-documentation.pdf] 2005.

61. Cerami EG, Bader GD, Gross BE, Sander C: cPath: open source software for collecting, storing, and querying biological pathways. BMC Bioinformatics 2006, 7:497.

62. Strömbäck L, Lambrix P: Representations of molecular pathways: an evaluation of SBML, PSI MI and BioPAX. Bioinformatics 2005, 21:4401–4407.

63. Brazma A, Krestyaninova M, Sarkans U: Standards for systems biology. Nat Rev Genet 2006, 7:593–605.

64. Davidson SB, Overton C, Buneman P: Challenges in integrating biological data sources. Journal of Computational Biology 1995, 2:557–572.

65. Zamboulis L, Martin N, Poulovassilis A: Bioinformatics Service Reconciliation By Heterogeneous Schema Transformation. Data Integration in the Life Sciences 2007, LNCS 2007, 4544:89–104.

66. Rahm E, Bernstein PA: A survey of approaches to automatic schema matching. VLDB J 2001, 10(4):334–350.

67. Laibe C, Le Novère N: MIRIAM Resources: tools to generate and resolve robust cross-references in Systems Biology. BMC Syst Biol 2007, 1:58.

68. Gruber TR: Towards Principles for the Design of Ontologies Used for Knowledge Sharing.   [http://citeseerx.ist.psu.edu/viewdoc/summary?doi=10.1.1.43.6200] In Formal Ontology in Conceptual Analysis and Knowledge Representation Edited by: Guarino N, Poli R. Deventer, The Netherlands: Kluwer Academic Publishers; 1993.

69. Fischer M, Thai QK, Grieb M, Pleiss J: DWARF – a data warehouse system for analyzing protein families. BMC Bioinformatics 2006, 7:495.

70. Lee TJ, Pouliot Y, Wagner V, Gupta P, Stringer-Calvert DW, Tenenbaum JD, Karp PD: BioWarehouse: a bioinformatics database warehouse toolkit. BMC Bioinformatics 2006, 7:170.

71. Shah SP, Huang Y, Xu T, Yuen MM, Ling J, Ouellette BF: Atlas – a data warehouse for integrative bioinformatics. BMC Bioinformatics 2005, 6:34.

72. Birkland A, Yona G: BIOZON: a system for unification, management and analysis of heterogeneous biological data. BMC Bioinformatics 2006, 7:70.

73. Shafer P, Isganitis T, Yona G: Hubs of knowledge: using the functional link structure in Biozon to mine for biologically significant entities. BMC Bioinformatics 2006, 7:71.

74. Wiederhold G: Mediators in the Architecture of Future Information Systems. IEEE Computer 1992, 25(3):38–49.

75. Pérez-Rey D, Maojo V, García-Remesal M, Alonso-Calvo R, Billhardt H, Martin-Sánchez F, Sousa A: ONTOFUSION: Ontology-based integration of genomic and clinical databases. [http://dx.doi.org/10.1016/j.compbiomed.2005.02.004] Comput Biol Med 2005.

76. Goble C, Stevens R, Ng G, Bechhofer S, Paton N, Baker P, Peim M, Brass A: Transparent Access to Multiple Bioinformatics Information Sources. IBM Systems Journal Special issue on deep computing for the life sciences 2001, 40(2):532–552.

77. Shaker R, Mork P, Brockenbrough JS, Donelson L, Tarczy-Hornoch P: The BioMediator System as a Tool for Integrating Biologic Databases on the Web. Proceedings of the VLDB 2004 Workshop on Information Integration on the Web 2004.

78. Wilkinson MD, Links M: BioMOBY: An Open Source Biological Web Services Proposal. Briefings in Bioinformatics 2002, 3(4):331–341.

79. Oinn TM, Greenwood RM, Addis M, Alpdemir MN, Ferris J, Glover K, Goble CA, Goderis A, Hull D, Marvin D, Li P, Lord PW, Pocock MR, Senger M, Stevens R, Wipat A, Wroe C: Taverna: lessons in creating a work flow environment for the life sciences. Concurrency and Computation: Practice and Experience 2006, 18(10):1067–1100.

80. Stevens RD, Robinson AJ, Goble CA: myGrid: personalised bioinformatics on the information grid. ISMB (Supplement of Bioinformatics) 2003, 302–304.

81. Delgado IN, del Mar Rojano-Muñoz M, Ramírez S, Pérez AJ, León EA, Montes JFA, Trelles O: Intelligent client for integrating bioinformatics services. Bioinformatics 2006, 22:106–111.

82. Gordon PMK, Sensen CW: Seahawk: moving beyond HTML in Web-based bioinformatics analysis. BMC Bioinformatics 2007, 8:208.

83. Carrere S, Gouzy J: REMORA: a pilot in the ocean of BioMoby web-services. Bioinformatics 2006, 22:900–901.

84. Kawas E, Senger M, Wilkinson MD: BioMoby extensions to the Taverna workflow management and enactment software. BMC Bioinformatics 2006, 7:523.

85. Shah AA, Barthel D, Lukasiak P, Blazewicz J, Krasnogor N: Web and Grid Technologies in Bioinformatics, Computational and Systems Biology: A Review. Current Bioinformatics 2008, 3:10–31.

86. Arenas M, Kantere V, Kementsietsidis A, Kiringa I, Miller RJ, Mylopoulos J: The hyperion project: from data integration to data coordination. SIGMOD Rec 2003, 32(3):53–58.

87. Kementsietsidis A, Arenas M, Miller RJ: Mapping Data in Peer-to-Peer Systems: Semantics and Algorithmic Issues. SIGMOD Conference 2003, 325–336.

88. Ooi BC, Tan KL, Zhou A, Goh CH, Li Y, Liau CY, Ling B, Ng WS, Shu Y, Wang X, Zhang M: PeerDB: peering into personal databases. Proceedings of the 2003 ACM SIGMOD international conference on Management of data, ACM 2003, 659–659.

89. Cudré-Mauroux P, Agarwal S, Aberer K: GridVine: An Infrastructure for Peer Information Management. IEEE Internet Computing 2007, 11(5):36–44.

90. Kementsietsidis A, Neven F, de Craen DV: BioScout: a life-science query monitoring system. EDBT 2008, 730–734.

91. Köhler J: Integration of life science databases. Drug Discovery Today: BIO-SILICO 2004, 2(2):61–69.

92. Shvaiko P, Euzenat J: Ten Challenges for Ontology Matching. Proc of ODBASE 2008, 1164–1182.

93. Jimenez-Ruiz E, Cuenca Grau B, Horrocks I, Berlanga R: Ontology Integration Using Mappings: Towards Getting the Right Logical Consequences. In Proc of European Semantic Web Conference (ESWC). Volume 5554. LNCS, Springer-Verlag; 2009:173–187.

94. Kalyanpur A, Parsia B, Sirin E, Grau BC: Repairing Unsatisfiable Concepts in OWL Ontologies. Proc of ESWC 2006, 170–184.

95. Meilicke C, Stuckenschmidt H, Tamilin A: Supporting Manual Mapping Revision using Logical Reasoning. Proc of AAAI 2008, 1213–1218.

96. Baader F: [http://www.amazon.ca/exec/obidos/redirect?tag=citeulike09–20-n&pa th=ASIN/0521781760] The Description Logic Handbook: Theory, Implementation and Applications. Cambridge University Press; 2003.

97. Neerincx PBT, Leunissen JAM: Evolution of web services in bioinformatics. Brief Bioinform 2005, 6:178–188.

98. Dibernardo M, Pottinger R, Wilkinson M: Semi-automatic web service composition for the life sciences using the BioMoby semantic web framework. J Biomed Inform 2008, 41:837–847.

99. Pérez JM, Llavori RB, Aramburu MJ, Pedersen TB: Integrating Data Warehouses with Web Data: A Survey. IEEE Trans Knowl Data Eng 2008, 20(7):940–955.

100. Polyzotis N, Garofalakis MN, Ioannidis YE: Approximate XML Query Answers. SIGMOD Conference 2004, 263–274.

101. Sanz I, Mesiti M, Guerrini G, Llavori RB: Fragment-based approximate retrieval in highly heterogeneous XML collections. Data Knowl Eng 2008, 64:266–293.

102. Adali S, Bonatti PA, Sapino ML, Subrahmanian VS: A Multi-Similarity Algebra. SIGMOD Conference 1998, 402–413.

103. Guerrini G, Mesiti M: X-Evolution: A Comprehensive Approach for XML Schema Evolution. Database and Expert Systems Applications, International Workshop on 2008251–255.

104. Moro MM, Malaika S, Lim L: Preserving XML queries during schema evolution. Proceedings of the 16th international conference on World Wide Web 2007, 1341–1342.

105. Velegrakis Y, Miller RJ, Popa L, Mylopoulos J: ToMAS: A System for Adapting Mappings while Schemas Evolve. Data Engineering, International Conference on 2004862.

106. Andritsos P, Fuxman A, Kementsietsidis A, Miller RJ, Velegrakis Y: Kanata: adaptation and evolution in data sharing systems. SIGMOD Rec 2004, 33(4): 32–37.

107. Haase P, Sure Y: D3.1.1.b State of the Art on Ontology Evolution. [Http://www.aifb.uni-karlsruhe.de/WBS/ysu/publications/SEKT-D3.1.1.b.pdf] Technical report 2004.

108. Yildiz B: Ontology Evolution and Versioning. The state of the art. [Http://publik.tuwien.ac.at/files/pub-inf_4603.pdf] Technical report 2006.

109. Hartung M, Kirsten T, Rahm E: Analyzing the Evolution of Life Science Ontologies and Mappings. DILS 2008, 11–27.

110. Castano S, Fugini MG, Martella G, Samarati P: Database security. ACM Press/Addison-Wesley Publishing Co; 1994.

111. Agrawal R, Srikant R: Privacy-preserving data mining. Proceedings of the 2000 ACM SIGMOD international conference on Management of data 2000, 439–450.

112. Verykios VS, Bertino E, Fovino IN, Provenza LP, Saygin Y, Theodoridis Y: State-of-the-art in privacy preserving data mining. SIGMOD Rec 2004, 33(1): 50–57.

113. Malin BA: An evaluation of the current state of genomic data privacy protection technology and a roadmap for the future. Journal of the American Medical Informatics Association 2005, 12(1):28–34.

114. Cios KJ, Moore GW: Uniqueness of medical data mining. Artificial Intelligence in Medicine 2002, 26:1–24.

115. Lupu E, Sloman M: Conflicts in policy-based distributed systems management. IEEE Transactions on Software Engineering 1999, 25:852–869.

116. Ahn GJ, Sandhu R: Role-based authorization constraints specification. ACM Trans Inf Syst Secur 2000, 3(4):207–226.

117. Sloman M, Lupu E: Security and management policy specification. IEEE Network 2002, 16:10–19.

118. Liu K, Ryan J: Random Projection-Based Multiplicative Data Perturbation for Privacy Preserving Distributed Data Mining. IEEE Trans on Knowl and Data Eng 2006, 18(1):92–106.

119. Sweeney L: k-anonymity: a model for protecting privacy. Int J Uncertain Fuzziness Knowl-Based Syst 2002, 10(5):557–570.

120. Rizvi S, Mendelzon A, Sudarshan S, Roy P: Extending query rewriting techniques for fine-grained access control. Proceedings of the 2004 ACM SIGMOD international conference on Management of data 2004, 551–562.

121. Williams A, Runte K: XML Format of the UniProt Knowledgbase. International Conference on Intelligent Systems for Molecular Biology 2004.

122. Mangalam H, et al.: GeneX: an open source gene expression database and integrated tool set. IBM Systems Journal 2001, 40(2):552–569.

123. Spellman P, et al.: Design and implementation of microarray gene expression markup language (MAGE-ML). Genome Biology 2002, 3(9):1–9.

124. Cuellar A, Nielsen P, Bullivant D, Hunter P: CellML 1.1 for the Definition and Exchange of Biological Models. CIFAC Symposium on Modelling and Control in Biomedical Systems 2003, 451–456.

125. Orchard S, Hermjakob H: The HUPO proteomics standards initiative-easing communication and minimizing data loss in a changing world. Briefings in Bioinformatics 2007, 9(2):166–173.

126. Murray-Rust P, Rzepa H: Chemical Markup, XML and the Worldwide Web. Part 4. CML Schema. J Chem Inf comp Sci 2003, 43:757–772.

Stringency of the 2-His–1-Asp Active-Site Motif in Prolyl 4-Hydroxylase

Kelly L. Gorres, Khian Hong Pua and Ronald T. Raines

ABSTRACT

The non-heme iron(II) dioxygenase family of enzymes contain a common 2-His–1-carboxylate iron-binding motif. These enzymes catalyze a wide variety of oxidative reactions, such as the hydroxylation of aliphatic C–H bonds. Prolyl 4-hydroxylase (P4H) is an α-ketoglutarate-dependent iron(II) dioxygenase that catalyzes the post-translational hydroxylation of proline residues in protocollagen strands, stabilizing the ensuing triple helix. Human P4H residues His412, Asp414, and His483 have been identified as an iron-coordinating 2-His–1-carboxylate motif. Enzymes that catalyze oxidative halogenation do so by a mechanism similar to that of P4H. These halogenases retain the active-site histidine residues, but the carboxylate ligand is replaced with a halide ion. We replaced Asp414 of P4H with alanine (to mimic the active site of a halogenase) and with glycine. These substitutions do not, however, *convert P4H into a halogenase. Moreover, the hydroxylase activity of D414A P4H cannot be rescued with small molecules. In addition, rearranging the two His and one Asp residues in the active site eliminates hydroxylase activity. Our results demonstrate a high stringency for the iron-binding residues in the P4H active site. We conclude that P4H, which catalyzes an especially demanding chemical transformation, is recalcitrant to change.*

Introduction

Iron is a common cofactor in enzymes that employ molecular oxygen as an oxidant. In addition to those iron-dependent enzymes that rely on heme, there is a class of non-heme, mononuclear iron(II) enzymes that catalyze the hydroxylation of aliphatic C–H bonds, dihydroxylation of arene double bonds, epoxidation of C–C double bonds, heterocyclic ring formation, and oxidative aromatic ring opening. The iron in these non-heme dioxygenases is commonly coordinated by a 2-His–1-carboxylate motif [1]. Two histidine (His) and either one aspartic acid (Asp) or one glutamic acid (Glu) residue bind the iron at the vertices of one triangular face of the octahedral metal center, forming a 2-His–1-carboxylate triad of facial ligands (Figure 1A). This arrangement leaves three coordination sites on the iron open for the substrate, oxygen, and other co-substrates.

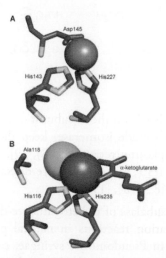

Figure 1. Active sites of α-ketoglutarate-dependent iron(II) dioxygenases. (A) The Chlamydomonas reinhardtii prolyl 4-hydroxylase contains a 2-His–1-carboxylate facial triad composed of His143, His227, and Asp145. A zinc ion (gray) replaced iron in the crystal structure [28]. (B) The SyrB2 halogenase has two histidine residues (His116, His235) that coordinate to the iron ion (orange). α-Ketoglutarate is also bound to the iron. Alanine (Ala118) is found in place of the carboxylate-containing residue of hydroxylases, allowing space for the chloride ion (green) [12].

A subset of non-heme iron(II) enzymes uses α-ketoglutarate as a co-substrate. These dioxygenases also contain the 2-His–1-carboxylate iron-binding motif. The substrate binds near the active site, while α-ketoglutarate binds directly to the iron in a bidentate manner. The reaction mechanism involves oxidative decarboxylation of α-ketoglutarate to produce succinate, CO_2, and a high energy Fe(IV)-oxo intermediate [2]. In a hydroxylation reaction, one atom of oxygen is incorporated into the product, and the other into succinate.

Prolyl 4-hydroxylase (P4H) is an α-ketoglutarate-dependent, iron(II) dioxygenase, and catalyzes the post-translational hydroxylation of proline residues (Pro) during collagen biosynthesis [3]. This hydroxylation reaction forms (2S,4R)-4-hydroxyproline (Hyp; Figure 2), which is necessary for the proper folding of the collagen triple helix [4]. Accordingly, P4H is an essential enzyme for animals [5]–[7].

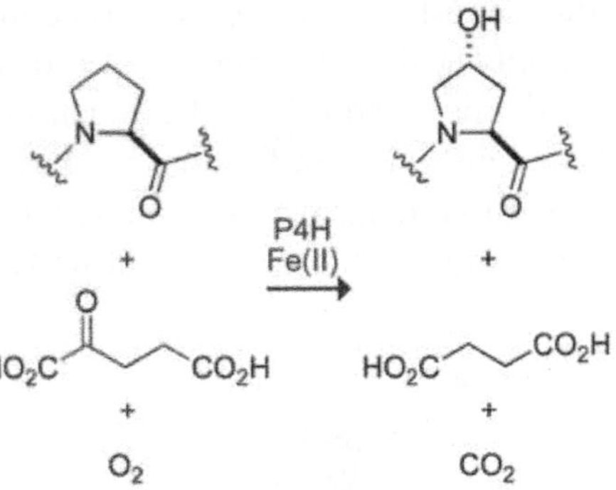

Figure 2. Reaction catalyzed by P4H.

The active site of P4H is within the α subunit of the α2β2 enzyme tetramer. The β subunits (protein disulfide isomerase) keep the α subunits soluble and retained within the endoplasmic reticulum [8], [9]. The 2-His–1-carboxylate iron-binding residues of P4Hα, identified by site-directed mutagenesis, comprise His412, Asp414, and His483 [10], [11].

A recently discovered subclass of α-ketoglutarate-dependent, iron(II) dioxygenases catalyzes halogenation reactions in natural product biosynthesis. The halogenase SyrB2, found in Pseudomonas syringae, catalyzes the conversion of threonine to 4-chlorothreonine during the biosynthesis of syringomycin E. The three-dimensional structure of SyrB2 revealed a chloride ion present in place of a carboxylate ligand in the facial triad [12]. The Asp or Glu present in other mononuclear iron enzymes is replaced by alanine (Ala) in SyrB2, creating space in the active site for the binding of a chloride ligand to the iron (Figure 1B).

The halogenase CytC3 likewise contains a 2-His motif rather than the 2-His–1-carboxylate facial triad [13].

Despite the difference in iron-binding residues, halogenation is thought to follow a mechanism similar to that of P4H and other α-ketoglutarate-dependent iron(II) dioxygenases (Figure 3). Decarboxylation of α-ketoglutarate generates a reactive Fe(IV)-oxo intermediate, which abstracts a hydrogen atom from a C–H bond of the substrate [14], [15]. This reaction yields a substrate radical and a Fe(III)-OH complex. Hydroxylated products result from the recombination of the substrate radical with the iron-coordinated hydroxyl group. In contrast, the radical substrate intermediate of halogenases combines with the chlorine atom coordinated to the iron, instead of the hydroxyl group, to produce a chlorinated product. This discovery prompts the question of how the enzyme controls product formation.

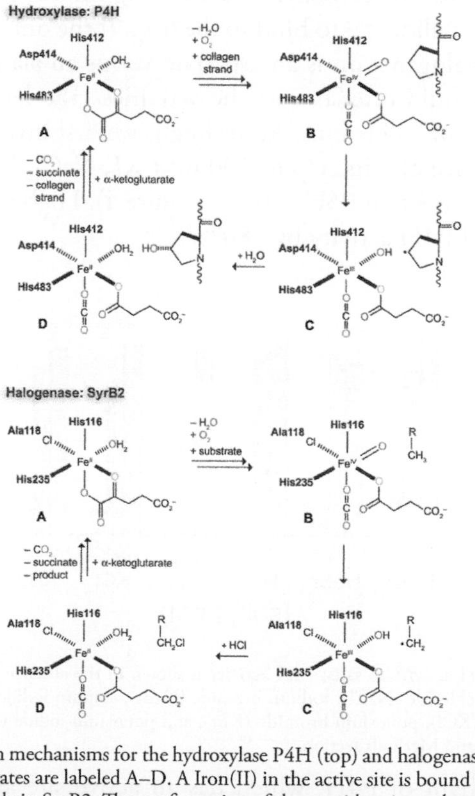

Figure 3. Proposed reaction mechanisms for the hydroxylase P4H (top) and halogenase SyrB2 (bottom). In both mechanisms, the intermediates are labeled A–D. A Iron(II) in the active site is bound by a 2-His–1-Asp motif in P4H, but a 2-His–1-chloride in SyrB2. The configuration of these residues is not known for human P4H, but is drawn in analogy to SyrB2 (Figure 2B). B The reactive iron(IV)-oxo species is formed upon decarboxylation of α-ketoglutarate. C The iron(IV)-oxo species abstracts a hydrogen atom from the substrate producing a radical intermediate. D In the hydroxylase reaction, the substrate radical recombines with the hydroxyl group. In the halogenase reaction, the substrate radical reacts with the chloride ligand.

Herein, we test the stringency of the iron-binding ligands in P4H. We determine whether altering the P4H active site so as to mimic that of a halogenase can convert its hydroxylase activity into halogenase activity. In addition, we characterize the spatial relationships within the active site of P4H by changing the relative positions of the residues that constitute the 2-His–1-carboxylate facial triad. The results provide insight on enzymes that catalyze difficult but important chemical transformations.

Results

Difference Between Hydroxylases and Halogenases

P4H, an α-ketoglutarate-dependent iron(II) dioxygenase, contains a 2-His–1-carboxylate motif. In related enzymes that catalyze halogenation reactions, instead of hydroxylations, the carboxylate-containing residue is replaced with Ala, providing space for a halide ion to bind to the iron. If the only difference between hydroxylation and halogenation is an aspartate versus an alanine, then exchanging these residues should interconvert the activities. The halogenation reaction requires the presence of halide ions. Accordingly, we first investigated the activity of P4H in the presence of salts. Up to 100 mM KF, NaCl, NaBr, NaI, KF, KCl, KBr, or KI had little effect on P4H activity (Figure 4). Higher salt concentrations (500 mM) decreased P4H activity by ~80%.

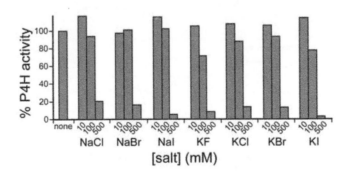

Figure 4. Tolerance of P4H activity to salts. P4H activity is shown in the absence and presence of increasing concentrations of sodium chloride (NaCl), sodium bromide (NaBr), sodium iodide (NaI), potassium fluoride (KF), potassium chloride (KCl), potassium bromide (KBr), and potassium iodide (KI). Reaction mixtures are described in the Materials and Methods section.

The D414A variant of P4H, which has an active site analogous to that of SyrB2, was produced, and, as expected [16], lost its hydroxylase activity. Still, no halogenated product from reaction mixtures containing D414A P4H was detected by high-performance liquid chromatography (HPLC) or mass spectrometry, even in the presence of 100 mM sodium fluoride, chloride, or bromide (Table 1). Subsequently, a D414G variant was made to provide even more space for a halide ion within the active site. This variant likewise had no detectable halogenase or hydroxylase activity.

Table 1. Biochemical attributes of P4H variants.

P4H variant	Poly(proline) affinity	Hydroxylase activity	Halogenase activity
Wild-type	+	+	−
D414A	+	−	−
D414G	+	−	−
D414H	+	−	nd
D414H/H412D	+	−	nd
D414H/H438D	+	−	nd

nd, Not determined.

Small Molecules Cannot Replace the Carboxylate in the P4H Active Site

Functional groups missing from amino-acid side chains can be restored by small molecules that rescue protein function [17]. In the D414A and D414G variants of P4H, the carboxylate group is absent. The sodium salt of azide, formate, or acetate (100 mM) was added to D414A P4H reaction mixtures, but did not lead to the formation of hydroxylated product. Nor was the activity of D414G rescued by nitrite or nitrate.

A 3-His Active Site Is Not Functional in P4H

A 3-His active site is found in cysteine dioxygenase, a non-heme, iron(II) dioxygenase related to P4H that does not utilize α-ketoglutarate. To mimic that enzyme, Asp414 was replaced with a histidine residue. The D414H variant was, however, unable to form a hydroxylated product detectable by either the HPLC-based enzyme assay or mass spectrometry.

Spatial Orientation of the 2-His–1-Asp Triad is Critical for Activity

The 2-His–1-Asp triad occupies adjacent coordination sites on the iron. The relative position of these residues varies with respect to the sites occupied by α-ketoglutarate and oxygen [18]. The locations of the two histidine and aspartate residues were shuffled in the H412D/D414H and D414H/H483D variants. No hydroxylated product was detected in reactions with either of these P4H variants by either HPLC or mass spectrometry. Apparently, the relative positions of the components of the 2-His–1-carboxylate triad are critical for the hydroxylation activity of P4H.

Discussion

The 2-His–1-carboxylate motif is common to many α-ketoglutarate-dependent iron(II) dioxygenases. The related halogenases differ in that the iron is coordinated by only two His residues. In the halogenases, Ala is located in place of the Asp/Glu, and a chloride ion binds to the iron in place of the carboxylate ligand. In an attempt to coerce P4H to perform halogenation reactions, we created the D414A and D414G variants of P4H. Although these variants lose their hydroxylase activity, they do not gain detectable halogenase activity. Similar experiments have been conducted in other dioxygenases to explore the conversion of hydroxylases to halogenases, and vice versa. Replacing alanine with aspartate or glutamate in the halogenase SyrB2 abolished halogenase activity with no observable hydroxylase activity [12]. Substitution of aspartate for alanine in hydroxylase taurine dioxygenase (TauD) abolished hydroxylase activity with no gain in halogenase activity [19]. The D101A variant of TauD did bind iron, though not as efficiently as did the wild-type enzyme, but it was not known if sufficient chloride ion was binding in the active site. The affinity of D414A P4H and D414G P4H for iron was not determined, as even the wild-type P4H does not bind iron with high affinity (K_M = 5 μM [20]). These P4H variants were able to be produced like the wild-type enzyme and had a similar affinity for poly(proline) (Table 1), as assessed by their retention during poly(proline)-affinity chromatography. Accordingly, we conclude that global protein conformation was unaffected by the amino-acid substitutions.

All data thus far shows that the difference between Fe(II) and α-ketoglutarate-dependent hydroxylases and halogenases extends beyond the presence or absence of an enzymic carboxyl group. Analysis of structural data of the halogenases CytC3 and SyrB2 suggests that residues that do not make a direct contact with the active-site iron are essential for binding α-ketoglutarate and a chloride ion [12], [13]. Residues that surround the active site form a network of hydrogen bonds, which appear to contribute to chloride binding in CytC3 and SyrB2, and could fail to do so in the P4H variants.

The hydroxylase activity could not be rescued by a number of small molecules. Chemical rescue of the residues within the 2-His–1-carboxylate motif of α-ketoglutarate-dependent iron(II) dioxygenases has been demonstrated with other enzymes. The activity of the H174A variant of phytanoyl-CoA 2-hydroxylase was rescued with the addition of imidazole [21]. Interestingly, the same substitution at the other iron-binding His residue was inert to rescue. Deleterious substitutions of the carboxylate ligand have also been rescued by small molecules, as demonstrated by the rescue of TauD D101A by formate [19]. These data provide clues about the accessibility of the enzymic active sites, as well as the tolerance to deviation in ligand position.

Although the 2-His–1-carboxylate motif is common to many iron(II) α-ketoglutarate-dependent dioxygenases, there are variations of the metal coordination motif [22]. Whereas the 2-His–1-carboxylate dioxygenases are quite selective for iron, enzymes containing three histidines and one glutamate (3-His–1-Glu) coordinate additional metals. Moreover, some dioxygenases do not utilize a carboxylate ligand at all. The halogenases, discussed previously, contribute just two histidine residues for iron binding. Enzymes with four histidine (4-His) and three histidine (3-His) ligands have also been identified. A comparison of the 3-His and 2-His–1-carboxylate metal sites revealed an overall structural similarity with slight differences in metal-ligand distances [23]. Conversion of a 2-His–1--carboxylate to a 3-His metal center was accomplished in tyrosine hydroxylase [24]. In P4H, the 3-His motif (that is, D414H P4H) did not have detectable enzymatic activity. Thus far, glutamate is the only amino acid that can replace Asp414 in P4H and maintain activity [11].

Instead of removing the carboxylate ligand, a second Asp/Glu could be added to the facial triad or the carboxylate could be relocated. His675 in aspartyl (asparaginyl) β-hydroxylase was replaced with Asp or Glu, and the enzyme retained 20 or 12% of its activity, respectively [25]. Gln or Glu could be substituted for His255 in TauD with 81 or 33% retention of its activity, respectively [19]. In P4H, however, no activity was found in the H412E or H483E variant [11]. Rearrangement of the His and Asp ligands in the P4H active site (as in the H412D/D414H and D414H/H483D variants) did not result in enzymatic activity, demonstrating that the native orientation of the 2-His–1-carboxylate facial triad is critical for P4H catalytic ability.

In conclusion, the 2-His–1-Asp facial triad in the active site of P4H is resistant to variation, including alternative iron-binding residues and the spatial positioning of the residues. Mutational analysis of the iron-binding residues within non-heme, iron(II) dioxygenases shows some flexibility at these locations, though no pattern has emerged as to which amino acids can endow function at a particular position. Although focus has been placed on the common 2-His–1-carboxylate iron-binding motif, differences among the non-heme iron(II) dioxygenases point toward the importance of residues that do not bind the iron directly. This secondary coordination sphere presumably alters the steric and electrostatic environments of the facial triad and influences the reaction mechanism. Three-dimensional structural information for human P4H, which has thus far eluded crystallization, will identify these additional residues, and could provide information that enables the design of P4H variants with transformed activity.

Materials and Methods

Materials

Escherichia coli strain Origami B (DE3) were obtained from Novagen (Madison, WI). Enzymes used for DNA manipulation were from Promega (Madison, WI) and DNA oligonucleotides for mutagenesis and sequencing were from Integrated DNA Technologies (Coralville, IA). DNA sequences were elucidated by capillary arrays on an Applied Bioscience automated sequencing instrument at the University of Wisconsin–Madison Biotechnology Center. Poly(proline) was from Sigma Chemical (St. Louis, MO). Luria–Bertani (LB) medium contained tryptone (10 g), yeast extract (5 g), and NaCl (10 g). Terrific broth (TB) medium contained tryptone (12 g), yeast extract (24 g), K_2HPO_4 (72 mM), KH_2PO_4 (17 mM), and glycerol (4 mL). All media were prepared in deionized, distilled water, and autoclaved.

Instrumentation

UV absorbance measurements were made with a Cary 50 spectrophotometer from Varian (Palo Alto, CA). Fast protein liquid chromatography (FPLC) was performed with an AKTA system from Amersham–Pharmacia (Piscataway, NJ) and the results were analyzed with the UNICORN Control System. High-performance liquid chromatography (HPLC) was carried out with a system from Waters (Milford, MA) that was controlled with the manufacturer's Millennium32 (Version 3.20) software. Mass spectrometry was performed with a Perkin–Elmer (Wellesley, MA) Voyager MALDI–TOF mass spectrometer at the University of Wisconsin–Madison Biophysics Instrumentation Facility.

Production and Purification of P4H Active–Site Variants

The pBK1.PDI1.P4H7 plasmid that directs the expression of the α and β subunits of human P4H served as a template for mutagenesis [26]. The QuikChange site-directed mutagenesis kit (Stratagene) was used to make point mutations in the gene encoding the α subunit. The P4H variants were produced and purified by using procedures reported previously [26]. Briefly, P4H cDNA expression in E. coli Origami B (DE3) cells was induced at 18°C for 18 h. Cells were harvested, lysed by sonication, and fractionated by precipitation with ammonium sulfate. P4H variants were purified by chromatography on a poly(proline)-affinity column, eluting with free poly(proline). Fractions were purified further by anion-exchange chromatography, followed by gel-filtration chromatography.

HPLC-Based Enzyme Activity Assay

The ability of P4H to catalyze the hydroxylation or halogenation of a proline residue was monitored with an HPLC-based assay described previously [16], [26], [27]. Assays were conducted for 20 min at 30°C in 100 μL of 50 mM Tris–HCl buffer, pH 7.8, containing dithiothreitol (100 μM), catalase (100 μg/mL), ascorbate (2 mM), FeSO4 (50–300 μM), α-ketoglutarate (0.5–25 mM), bovine serum albumin (1.0 mg/mL), and P4H (0.09–1.5 μM). The synthetic tetrapeptide substrate (dansyl-Gly–Phe–Pro–GlyOEt) was added to initiate the reaction. A reversed-phase analytical C18 column was used to separate the substrate and product peptides. Enzymic reaction mixtures were also analyzed by mass spectrometry. To assess the effect of salt on catalysis, KF, NaCl, NaBr, NaI, KF, KCl, KBr, or KI was added to reaction mixtures to 10, 100 or 500 mM. To attempt to rescue the enzymatic activity of D414A P4H and D414G P4H, the sodium salt of azide, nitrite, nitrate, formate, or acetate was added to reaction mixtures to 100 mM.

Authors' Contributions

Conceived and designed the experiments: KLG RTR. Performed the experiments: KLG KHP. Analyzed the data: KLG KHP RTR. Wrote the paper: KLG RTR.

References

1. Koehntop KD, Emerson JP, Que L, Jr. (2005) The 2-His–1-carboxylate facial triad: A versatile platform for dioxygen activation by mononuclear non-heme iron(II) enzymes. J Biol Inorg Chem 10: 87–93.

2. Costas M, Mehn MP, Jensen MP, Que L, Jr. (2004) Dioxygen activation at mononuclear nonheme iron active sites: Enzymes, models, and intermediates. Chem Rev 104: 939–986.

3. Myllyharju J (2003) Prolyl 4-hydroxylases, the key enzymes of collagen biosynthesis. Matrix Biol 22: 15–24.

4. Shoulders MD, Raines RT (2009) Collagen structure and stability. Annu Rev Biochem 78: 929–958.

5. Winter AD, Page AP (2000) Prolyl 4-hydroxylase is an essential procollagen-modifying enzyme required for exoskeleton formation and the maintenance of body shape in the nematode Caenorhabditis elegans. Mol Cell Biol 20: 4084–4093.

6. Friedman L, Higgin JJ, Moulder G, Barstead R, Raines RT, et al. (2000) Prolyl 4-hydroxylase is required for viability and morphogenesis in Caenorhabditis elegans. Proc Natl Acad Sci USA 97: 4736–4741.

7. Holster T, Pakkanen O, Soininen R, Sormunen R, Nokelainen M, et al. (2007) Loss of assembly of the main basement membrane collagen, Type IV, but not fibril-forming collagens and embryonic death in collagen prolyl 4-hydroxylase I null mice. J Biol Chem 282: 2512–2519.

8. Vuori K, Pihlajaniemi T, Myllyla R, Kivirikko KI (1992) Site-directed mutagenesis of human protein disulphide isomerase: Effect on the assembly, activity and endoplasmic reticulum retention of human prolyl 4-hydroxylase in Spodoptera frugiperda insect cells. EMBO J 11: 4213–4217.

9. Vuori K, Pihlajaniemi T, Marttila M, Kivirikko KI (1992) Characterization of the human prolyl 4-hydroxylase tetramer and its multifunctional protein disulfide-isomerase subunit synthesized in a baculovirus expression system. Proc Natl Acad Sci USA 89: 7467–7470.

10. Lamberg A, Helaakoski T, Myllyharju J, Peltonen S, Notbohm H, et al. (1996) Characterization of human type III collagen expressed in a baculovirus system. J Biol Chem 271: 11988–11995.

11. Myllyharju J, Kivirikko KI (1997) Characterization of the iron- and 2-oxoglutarate-binding sites of human prolyl 4-hydroxylase. EMBO J 16: 1173–1180.

12. Blasiak LC, Vaillancourt FH, Walsh CT, Drennan CL (2006) Crystal structure of the non-haem iron halogenase SyrB2 in syringomycin biosynthesis. Nature 440: 368–371.

13. Wong C, Fujimori DG, Walsh CT, Drennan CL (2009) Structural analysis of an open active site conformation of nonheme iron halogenase CytC3. J Am Chem Soc 131: 4872–4879.

14. Hoffart LM, Barr EW, Guyer RB, Bollinger JM, Jr., Krebs C (2006) Direct spectroscopic detection of a C–H-cleaving high-spin Fe(IV) complex in a prolyl-4-hydroxylase. Proc Natl Acad Sci USA 103: 14738–14743.

15. Galonic DP, Barr EW, Walsh CT, Bollinger J, J.M., Krebs C (2007) Two interconverting Fe(IV) intermediates in aliphatic chlorination by the halogenase CytC3. Nat Chem Biol 3: 113–116.

16. Gorres KL, Raines RT (2009) Direct and continuous assay for prolyl 4-hydroxylase. Anal Biochem 386: 181–185.

17. Qiao Y, Molina H, Pandey A, Zhang J, Cole PA (2006) Chemical rescue of a mutant enzyme in living cells. Science 311: 1293–1297.

18. Clifton IJ, McDonough MA, Ehrismann D, Kershaw NJ, Granatino N, et al. (2006) Structural studies on 2-oxoglutarate oxygenases and related double-stranded β-helix fold proteins. J Inorg Biochem 100: 644–669.

19. Grzyska PK, Muller TA, Campbell MG, Hausinger RP (2007) Metal ligand substitution and evidence for quinone formation in taurine/α-ketoglutarate dioxygenase. J Inorg Biochem 101: 797–808.

20. Tuderman L, Myllyla R, Kivirikko KI (1977) Mechanism of the prolyl hydroxylase reaction. 1. Role of co-substrates. Eur J Biochem 80: 341–348.

21. Searls T, Butler D, Chien W, Mukherji M, Lloyd MD, et al. (2005) Studies on the specificity of unprocessed and mature forms of phytanoyl-CoA 2-hydroxylase and mutation of the iron binding ligands. J Lipid Res 46: 1660–1667.

22. Straganz GD, Nidetzky B (2006) Variations of the 2-His–1-carboxylate theme in mononuclear non-heme FeII oxygenases. ChemBioChem 7: 1536–1548.

23. Leitgeb S, Nidetzky B (2008) Structural and functional comparison of 2-His–1-carboxylate and 3-His metallocentres in non-haem iron(II)-dependent enzymes. Biochem Soc Trans 36: 1180–1186.

24. Fitzpatrick PF, Ralph EC, Ellis HR, Willmon OJ, Daubner SC (2003) Characterization of metal ligand mutants of tyrosine hydroxylase: Insights into the plasticity of a 2-histidine-1-carboxylate triad. Biochemistry 42: 2081–2088.

25. McGinnis K, Ku GM, VanDusen WJ, Fu J, Garsky V, et al. (1996) Site-directed mutagenesis of residues in a conserved region of bovine aspartyl (asparaginyl) β-hydroxylase: Evidence that histidine 675 has a role in binding Fe^{2+}. Biochemistry 35: 3957–3962.

26. Kersteen EA, Higgin JJ, Raines RT (2004) Production of human prolyl 4-hydroxylase in Escherichia coli. Protein Exp Purif 38: 279–291.

27. Gorres KL, Edupuganti R, Krow GR, Raines RT (2008) Conformational preferences of substrates for human prolyl 4-hydroxylase. Biochemistry 47: 9447–9455.

28. Koski MK, Hieta R, Hirsila M, Ronka A, Myllyharju J, et al. (2009) The crystal structure of an algal prolyl 4-hydroxylase complexed with a proline-rich peptide reveals a novel buried tripeptide binding motif. J Biol Chem 284: 25290–25301.

Biosynthesis of the Proteasome Inhibitor Syringolin A: The Ureido Group Joining Two Amino Acids Originates from Bicarbonate

Christina Ramel, Micha Tobler, Martin Meyer, Laurent Bigler, Marc-Olivier Ebert, Barbara Schellenberg and Robert Dudler

ABSTRACT

Background

Syringolin A, an important virulence factor in the interaction of the phyto-pathogenic bacterium Pseudomonas syringae pv. syringae B728a with its host plant Phaseolus vulgaris (bean), was recently shown to irreversibly inhibit eu-karyotic proteasomes by a novel mechanism. Syringolin A is synthesized by a mixed non-ribosomal peptide synthetase/polyketide synthetase and consists of a tripeptide part including a twelve-membered ring with an N-terminal va-line that is joined to a second valine via a very unusual ureido group. Analy-sis of sequence and architecture of the syringolin A synthetase gene cluster with the five open reading frames sylA-sylE allowed to formulate a biosynthesis model that explained all structural features of the tripeptide part of syringolin A but left the biosynthesis of the unusual ureido group unaccounted for.

Results

We have cloned a 22 kb genomic fragment containing the sylA-sylE gene clus-ter but no other complete gene into the broad host range cosmid pLAFR3. Transfer of the recombinant cosmid into Pseudomonas putida and P. syrin-gae pv. syringae SM was sufficient to direct the biosynthesis of bona fide sy-ringolin A in these heterologous organisms whose genomes do not contain ho-mologous genes. NMR analysis of syringolin A isolated from cultures grown in the presence of NaH^{13}CO$_3$ revealed preferential ^{13}C-labeling at the ureido carbonyl position.

Conclusion

The results show that no additional syringolin A-specific genes were needed for the biosynthesis of the enigmatic ureido group joining two amino acids. They reveal the source of the ureido carbonyl group to be bicarbonate/carbon diox-ide, which we hypothesize is incorporated by carbamylation of valine medi-ated by the sylC gene product(s). A similar mechanism may also play a role in the biosynthesis of other ureido-group-containing NRPS products known largely from cyanobacteria.

Background

Syringolins are a family of closely related cyclic peptide derivatives that are se-creted by many strains of the phytopathogenic bacterium Pseudomonas syringae pv. syringae (Pss) in planta and under certain culture conditions [1,2]. Syringolin A, the major variant, was shown not only to induce acquired resistance in rice and wheat after spray application, but also to trigger hypersensitive cell death at infection sites of wheat and Arabidopsis plants infected by compatible powdery mildew fungi [3,4]. Recently, syringolin A was shown to be an important viru-lence factor in the interaction of Pss B728a with its host plant Phaseolus vulgaris (bean), and its cellular target has been identified. Syringolin A irreversibly inhibits the eukaryotic proteasome by a novel mechanism, representing a new structural class of proteasome inhibitors [5,6].

Structure elucidation revealed that syringolin A is a tripeptide derivative con-sisting of an N-terminal valine followed by the two non-proteinogenic amino acids 3,4-dehydrolysine and 5-methyl-4-amino-2-hexenoic acid, the latter two forming a twelve-membered macrolactam ring. The N-terminal valine is in turn linked to a second valine via an unusual ureido group (Figure 1A; [1]). The minor variants syringolin B to syringolin F differ from syringolin A by the substitution of one or both valines with isoleucine residues, by the substitution of 3,4-dehydro-lysine with lysine, and by combinations thereof [2]. The structure of syringolin A suggested that it was synthesized by a non-ribosomal peptide synthetase (NRPS), large modular enzymes that activate and condense amino acids according to the thiotemplate mechanism (for reviews see e.g. [7-9]). We previously cloned and delimited by mutational analysis a genomic region from Pss B301D-R containing five open reading frames (sylA-sylE) necessary for syringolin biosynthesis (Figure 1B; [10]). Whereas sylA and sylE encode a putative transcription activator and an exporter, respectively, sylC encodes a typical NRPS module predicted to activate valine, whereas sylD codes for two additional NRPS modules (of which the first is predicted to activate lysine and the second is predicted to activate valine [10]) and a type I polyketide synthetase (PKS) module. Type I PKS are also modular enzymes that, similar to fatty acid synthesis, extend a starter molecule by condensation/decarboxylation of malonate extender units (for reviews see e.g. [11,12]). The analysis of the structure and architecture of the syl gene cluster led to the postulation of a model that completely accounts for the biosynthesis of the tripeptide part of syringolin A, including its ring structure with the 5-methyl-4--amino-2-hexenoic acid and the 3,4-dehydrolysine (Figure 1C, [10]). However, although the addition of the ureido group and its attached second valine could not be explained by the model, the syl gene cluster did not contain additional open reading frames, which, if present, could potentially have been involved in the biosynthesis of this unexplained part.

Here we show that the genes sylA-sylE are sufficient to direct the biosynthesis of bona fide syringolin A when heterologously expressed in Pseudomonas putida and Pss SM, two organisms which do not produce syringolin A and have no syl gene cluster homolog in their genomes. Thus, biosynthesis of the ureido group with its attached terminal valine is achieved without additional syringolin A-specific genes (i. e. genes with no other function than in syringolin A biosynthesis). We hypothesized that biosynthesis of the ureido group would most likely be accomplished by the product of the sylC gene, which would, in addition to the extracyclic peptidyl valine, also activate the terminal valine and join the two residues by incorporation of a carbonyl group derived from hydrogen carbonate/carbon dioxide, thus forming the ureido moiety. We demonstrate by NMR spectroscopic analysis of syringolin A isolated from Pss cultures grown in the presence of $NaH^{13}CO_3$ that the ^{13}C isotope is preferentially found at the position of the ureido carbonyl atom. These results support our hypothesis, which may be of relevance for the hitherto unknown biosynthesis of other ureido-group-containing NRPS products largely known to be produced by cyanobacteria [13–20].

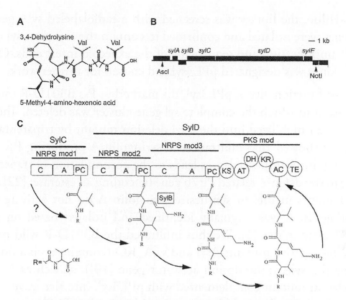

Figure 1. Structure and biosynthesis model of syringolin A. A. Structure of syringolin A. Amino acid constituents are delimited by bars. Val, valine. B. Genomic region of Pss B301D-R containing the sylA-sylE genes. Boxes above and below the line denote ORFs on the top and the bottom strand, respectively. Arrows indicate restriction sites used for cloning of the gene cluster into the cosmid pPL3syl. The sylA, sylB, and sylE genes encode a LuxR-type transcription activator, a rhizobitoxin desaturase-like protein thought to desaturate the lysine residue, and an efflux transporter, respectively [10]. The sylC gene encodes an NRPS module, while sylD codes for two NRPS modules and one PKS module [10] C. Biosynthesis model of the tripeptide part of syringolin A. The open boxes represent domains in modules of the syringolin A synthetase labeled with C, condensation domain; A, adenylation domain; PC, peptide carrier protein; KS, ketoacyl synthase; AT, acyl transferase; DH, dehydratase; KR, ketoreductase; AC, acyl carrier protein; TE, thioesterase. The A domains of the NRPS modules are thought to activate valine (NRPS mod1), lysine (NRPS mod2), and valine (NRPS mod3) [10]. The question mark indicates the unexplained synthesis and attachment of this group. The figures are adapted from [10].

Results

Biosynthesis of Syringolin A in Heterologous Organisms

In order to test whether the sylA-sylE gene cluster was sufficient to direct syringolin A biosynthesis, we constructed a cosmid containing the sylA-sylE genes but no other complete gene by taking advantage of AscI and NotI restriction sites flanking the syl gene cluster (Figure 1B). Southern blot analysis of AscI/NotI-digested genomic DNA of Pss B301D-R probed with a sylA gene fragment labeled the expected 22 kb fragment and thus confirmed the uniqueness of the restriction sites in the relevant genome region (data not shown). Thus, Pss B301D-R genomic DNA digested with AscI and NotI was separated by agarose gel electrophoresis. Fragments in the 20-23 kb size range were eluted and cloned into the wide host range cosmid pLAFR3 [21]. After packaging into lambda phages and transfection into

E. coli XL-1Blue, the library was screened with a radiolabeled sylA gene probe. Positive clones were isolated and confirmed to contain the complete syl gene cluster by PCR amplification and sequencing of the expected insert ends. One of the confirmed clones was designated pPL3syl and chosen for further work.

To test the functionality of pPL3syl, the markerless Pss B301D-R mutant Δsyl was constructed in which the complete syl gene cluster was deleted. The pPL3syl cosmid was then mobilized into the Δsyl deletion mutant by triparental mating. We previously showed that infiltration of syringolin A-producing Pss strains or isolated syringolin A into rice leaves leads to the accumulation of transcripts corresponding to the defense-related Pir7b gene (encoding an esterase; [22]), whereas strains or mutants unable to synthesize syringolin A do not activate this gene [1,23]. Syringolin A was originally identified and isolated based on its action on the Pir7b gene in rice [1]. We thus infiltrated the B301D-R wild-type strain, the syringolin-negative mutants Δsyl and sylA_KO (contains a plasmid insertion interrupting the sylA transcription activator gene [10]), as well as Δsyl (pPL3-syl), the deletion mutant complemented with pPL3syl, into rice leaves. RNA was extracted and subjected to gel blot analysis with regard to Pir7b transcript accumulation. As expected and in contrast to the wild type, the syringolin A-negative mutants did not induce Pir7b transcript accumulation, whereas the deletion mutant complemented with the pPL3syl cosmid led to a much stronger induction of the Pir7b gene (Figure 2A). This strongly suggested that pPL3syl contained a functional syl gene cluster able to direct syringolin A synthesis in the Δsyl deletion mutant. This does not exclude the possibility that genes not present in the syl gene cluster are necessary for syringolin A production because such genes would also be present in the Δsyl mutant background.

Next we wanted to mobilize pPL3syl into Pseudomonas strains not carrying syl gene homologs and lacking syringolin A production as evidenced by PCR, DNA gel blot analysis of genomic DNA, high performance liquid chromatography (HPLC) analysis of culture supernatants with regard to syringolin A content, infiltration into rice leaves followed by monitoring of Pir7b transcript accumulation, and whole genome sequence comparisons where possible (data not shown). After repeated unsuccessful attempts to transfer pPL3syl into the P. syringae pv. tomato DC3000 strain (all tetracycline-resistant putative transformants analyzed contained deletion variants of pPL3syl), the cosmid was successfully transferred into the non-pathogenic bacterium P. putida P3 [24] and Pss SM, a strain originally isolated from wheat [23,25]. Gel blot analysis of RNA extracted from rice leaves infiltrated with parental and transformed strains showed that, as expected, P. putida P3 and Pss SM did not induce Pir7b transcript accumulation. In contrast, both strains lead to Pir7b gene induction when carrying the pPL3syl cosmid (Figure 2B), suggesting that pPL3syl conferred the ability for syringolin A biosynthesis to these strains.

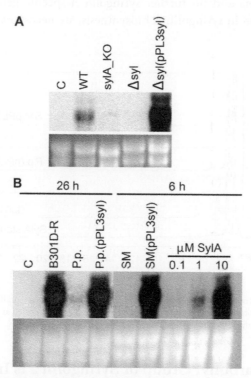

Figure 2. Gel blot analysis of Pir7b transcript accumulation. A. RNA was extracted from rice leaves infiltrated with water (C), Pss B301D-R (WT), a sylA plasmid insertion mutant (sylA_KO), a syl cluster deletion mutant (Δsyl), and Δsyl (pPL3syl), the deletion mutant complemented with the wild-type syl gene cluster. Top panel, autoradiogram (exposed for 5 h); bottom panel, ethidium bromide (EtBr)-stained agarose gel. B. Lanes were loaded with RNA extracted from rice leaves infiltrated as indicated. C, water control; B301D-R, Pss wild-type strain; P.p, P. putida P3; P.p. (pPL3syl), P. putida P3 transformed with the syl gene cluster; SM, Pss SM; SM (pPL3syl); Pss SM transformed with pPL3syl, and syringolin A solutions of the indicated concentrations. Top panel, autoradiogram (exposure times indicated on top), bottom panel, EtBr-stained gel.

To confirm this, the transformed strains were grown in shaken cultures in SRMAF medium and conditioned media were analyzed by HPLC. As shown in Figure 3, both strains produced a compound eluting at 15.5 min, the elution time of the syringolin A standard. Peaks were collected from multiple HPLC runs and subjected to mass spectrometry. HPLC-high resolution-electrospray ionization-mass spectrometry (HPLC-HR-ESI-MS) of the peaks from Pss SM and P. putida P3 carrying pPL3syl, and the Pss B301D-R wild type showed quasi-molecular ions [M+H]+ at m/z 494.29808 (1.5 ppm difference from calculated exact mass), 494.29653 (1.1 ppm), and 494.29799 (1.4 ppm), respectively, matching the empirical formula $C_{24}H_{40}N_5O_6+$ (protonated adduct of syringolin A; calculated exact mass 494.29731). We conclude from these experiments that the syl genes contained in pPL3syl are sufficient to direct syringolin A biosynthesis in these heterologous strains and no further syringolin A-specific genes, i.e. genes that exclusively function in syringolin A biosynthesis, are necessary.

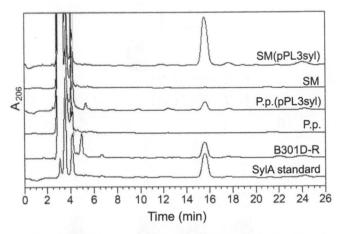

Figure 3. HPLC analysis of syringolin A content in conditioned SRM$_{AF}$ media. Conditioned media were sterile-filtrated and 20-µl-aliquots were loaded on the column. Absorption was monitored at 206 nm. Labels of HPLC traces are the same as in the legend to Figure 2B.

The Ureido Carbonyl Group of Syringolin a is Incorporated From Bicarbonate/Carbon Dioxide

The above results raised the question of how the ureido-valine is synthesized and incorporated into syringolin A. We hypothesized that this would most likely be accomplished by the product of the sylC gene, which would, in addition to the N-terminal valine of the tripeptide part of syringolin A, also activate the second valine and join the two residues via their amino groups formally by amidation of carbonic acid, thus forming the ureido moiety. If true, feeding syringolin A-producing cultures with ^{13}C-labeled hydrogen carbonate should result in syringolin A that is preferentially labeled with ^{13}C at the ureido carbonyl position. Thus, Pss B301D-R transformed with pOEAC, a plasmid carrying the sylA transcriptional activator gene under the control of the lacZ promoter, was grown in SRM$_{AF}$ medium. After 48 h, ^{13}C-labeled sodium hydrogen carbonate was added to a final concentration of 70 mM and the culture was further grown for 20 h. Syringolin A was isolated from conditioned medium as described [4] and subjected to ^{13}C NMR analysis.

The spectrum of labeled syringolin A was normalized in order to get the same signal intensities for the valine methyl groups as in the unlabeled sample. Comparison of the normalized NMR spectra revealed that the signal from the ureido carbon atom in 13C-labeled syringolin A was 45-fold stronger than the corresponding signal from unlabeled syringolin A (Figure 4). Inspection of the resolved 13C satellite of the valine methyl group at lowest field in the 1H spectrum of labeled syringolin A (data not shown) suggests a 13C content close to natural abundance. Therefore, the 45-fold signal enhancement in labeled syringolin A directly corresponds to the absolute 13C enrichment at this site. The normalized signal strengths of all other C atoms were equal in labeled and unlabeled syringolin A, with the exception of the C4 position of 3,4-dehydrolysine, whose signal was enhanced approximately 16-fold in 13C-labeled syringolin A (Figure 4A, B). Inspection of biosynthetic pathways using the KEGG database [26] revealed that this can be attributed to a carboxylation reaction in the biosynthesis of lysine. The C4 atom of lysine represents the C4 atom of L-aspartate-4-semialdehyde, a derivative of aspartate, which is condensed to pyruvate to yield the intermediary compound L-2,3-dihydrodipicolinate in bacterial lysine biosynthesis. The C4 atom of aspartate in turn originates from the carboxylation of pyruvate to oxaloacetate, an intermediary compound in the tricarboxylic acid cycle, which is transaminated to aspartate. Thus, enhanced 13C-labeling of lysine with $H13CO3-$ at the C4 position is to be expected. We note that malonate will also be labeled by $H13CO3-$ as it is derived from acetate by carboxylation. However, the label will be removed by the condensation/decarboxylation of malonate to the peptide chain during syringolin A biosynthesis. We conclude from this analysis that our hypothesis is correct, i.e. that the ureido carbonyl moiety in syringolin A originates from the incorporation of hydrogen carbonate/carbon dioxide.

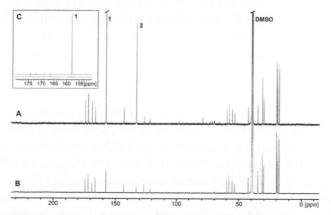

Figure 4. ^{13}C-NMR spectra of in vivo $NaH^{13}CO_3$-labeled and unlabeled syringolin A. The spectra of $NaH^{13}CO_3$-labeled (A) and unlabeled (B) syringolin A have been scaled to give equal signal intensities for the methyl groups of the valine residues (17.4, 17.5, 19.0, and 19.1 ppm shifts). The signals at 157.8 ppm (marked 1; clipped off) and 132.8 ppm (marked 2) correspond to the ureido carbonyl group and the lysine C4 position, respectively. DMSO, DMSO solvent signal. C. Scaled-down version of part of the spectra given in (A) and (B) to show the difference in signal intensity of the ureido carbonyl group in $NaH^{13}CO_3$-labeled and unlabeled syringolin A, respectively.

Discussion

We have demonstrated that the syl gene cluster is sufficient to direct syringolin A synthesis in heterologous organisms. Although the biosynthesis model presented earlier [10] plausibly explained every structural feature of the syringolin A tripeptide part through the enzymatic actions of the sylB, sylC, and sylD gene products, the generation and condensation of the ureido valine remained enigmatic. As the other genes present in the syl cluster encode a transcriptional activator (sylA gene) and an exporter (sylE gene), a plausible hypothesis was that the sylC-encoded NRPS module not only activated the N-terminal peptidyl valine, but also the ureido valine, and that the ureido carbonyl moiety is incorporated from hydrogen carbonate/carbon dioxide. As shown above, in vivo labeling of syringolin A with ^{13}C-hydrogen carbonate supports this hypothesis. Currently, we can only speculate how this is achieved. One possibility is that the quaternary syringolin A synthetase complex may contain two (not necessarily identical) molecules derived from the sylC gene per SylD polypeptide. Both sylC gene products would activate valine, or, to a certain degree, isoleucine in minor syringolin variants [2]. The first valine would then be carbamylated by HCO_3^-/CO_2, perhaps without the action of another enzyme, as has been reported for the carbamylation of a catalytic lysine residue in β-lactamases of class D [27,28]. The ureido moiety would then be formed by amide bond formation between the carbamylated valine and the second valine. In this scenario, it remains unclear how the first valine, which, like the second one, is envisioned to be bound to the peptide carrier protein domain by a thioester bond, is released upon ureido bond formation. It is also conceivable that ureido bond formation is achieved by a single SylC protein, which contains a condensation domain usually absent from starter modules that may be involved. To clarify these issues, more structural information about the large syringolin A synthetase and the SylC module must be obtained. The reconstitution of the enzymatic activities of the module(s) derived from the sylC gene in vitro will be challenging.

In addition to the syringolin family of compounds, a number of other natural cyclic peptides mostly isolated from cyanobacteria have been described in the literature that contain extracyclic ureido groups linking two different amino acids. These include anabaenopeptins from Anabaena, Oscillatoria, and Planktothrix species [13-16], ferintoic acids from Mycrocystis aeruginosa [17], pompanopeptins from Lyngbya confervoides [18], as well as mozamides and brunsvicamides, compounds of presumably cyanobacterial origin isolated from sponges [19,20]. Bicarbonate/CO2 may also be the source of the ureido carbonyl group joining two extracyclic amino acids in the biosynthesis of these compounds, which, to our knowledge, has not been elucidated so far.

Conclusion

Our results show that the syl biosynthesis gene cluster was sufficient to direct the biosynthesis of bona fide syringolin A, including the enigmatic ureido group joining two amino acids. They reveal the source of the ureido carbonyl group to be bicarbonate/carbon dioxide, which we hypothesize is incorporated by carbamylation of valine mediated by the sylC gene product(s). A similar mechanism may also play a role in the biosynthesis of other ureido-group-containing NRPS products known largely from cyanobacteria.

Methods

Construction and Expression of pPL3syl

Unless stated otherwise, standard protocols were used [29]. Genomic DNA from Pss B301D-R was isolated and 11 μg were digested with the restriction enzymes AscI and NotI. AscI and NotI sites are both unique in the syl gene region (Gen-Bank: AJ548826) located at position 2052 and 24124, respectively, within the ORFs flanking the sylA-sylE ORFs (3507-23596). A DNA gel blot was prepared with 1 μg of the digested DNA and probed with a ^{32}P-labeled sylA gene fragment PCR-amplified from genomic DNA with primers P1 (5'-ccatcgatggagtagagtgatg-gc) and P2 (5'-ggaattcttacaaaattcccatcttg). The rest of the digested DNA was separated on a 0.4% agarose gel and the DNA in the 20-23 kb size range was cut out, electrophoretically eluted into a dialysis bag (10 kDa cutoff), extracted first with 1 volume of phenol and then with 1 volume of phenol-chloroform-isoamylalcohol (25:24:1), precipitated with ethanol and finally taken up in TE (10 mM Tris-HCl, pH 8; 1 mM EDTA). Fragments were ligated into the HindIII/BamHI-cut broad host range cosmid vector pLAFR3 [21] using adaptors prepared by annealing the oligonucleotide 5'-cgcgccaagcttcca with 5'-agcttggaaagcttgg (AscI/HindIII adaptor) and 5'-ggccgctagtcaggag with 5'-gatcctcctgactagc (NotI/BamHI adaptor), respectively. Ligation products were packaged into lambda phage particles using the Gigapack III Gold Packaging Kit (Stratagene, La Jolla, California) and the library was plated out on E. coli XL1-Blue (Stratagene) and screened according to the instructions of the manufacturer using the ^{32}P-labeled sylA gene fragment described above as a probe. Positive clones were isolated and confirmed to contain the complete syl gene cluster by PCR amplification and sequencing of the insert end fragments using primers 5'-ccggcctacacgcattc (sylA end) and 5'-agcaacctggat-gtacgg (sylE end) with the respective adaptor oligonucleotides (see above).

pPL3syl was transferred from XL1-Blue to Pseudomonas strains by triparental mating using the E. coli helper strain HB101 (pRK600) [30,31].

Construction of the Syl Gene Cluster Deletion Mutant Δsyl

Two fragments of 783 bp and 655 bp length flanking the syl gene cluster on the 5' and 3' side, respectively, were amplified by PCR from Pss B301D-R genomic DNA using the primer pairs P3 (5'-cgggatccaacctgaaatgggagagtc; base given in bold at position 2297 in GenBank:AJ548826) and P4 (5'-agcgcgaggactcaatgt-gaaaacaacg; bold base at position 3072), and P5 (5'-tcacattgagtcctcgcgctggtaacc; bold base at position 23600) and P6 (5'-ttctgcagtcaagcctgacgaaaagc; bold base at position 24247), respectively. The two bands were isolated and joined by overlap extension PCR using primers P3 and P6 to yield a fragment flanked by BamHI and PstI restriction sites in which the syl gene cluster from position 3073-23599 (GenBank:AJ548826) was missing. The deletion is nearly identical with the one in the completely sequenced P. syringae pv. tomato DC3000 (GenBank:NC_004578.1), which does not contain a syl gene cluster. The fragment was cut with BamHI and PstI and cloned into the respective restriction sites in the cloning box of the suicide vector pME3087 (Tc^R, ColE1 replicon [32]). The recombinant plasmid was transformed into E. coli S17-1 (thi pro hsdR recA; chromosomal RP4 (Tra$^+$ Tc^S Km^S Ap^S; transfer gene-positive, tetracycline-sensitive, kanamycin-sensitive, ampicillin-sensitive) [33]) and mobilized into Pss B301D-R. Tetracycline-resistant colonies were grown in LB medium over night at 28°C on a rotary shaker (220 rpm). For selection of tetracycline-sensitive colonies, the overnight cultures were diluted 100-fold with LB. After 2 h of growth, tetra-cycline was added (20 µg/ml final concentration) and the cultures were grown for 1 h, after which the bactericide carbenicillin (2 mg/ml final concentration) was added for 3 h. The bacteria were then collected by centrifugation, and after wash-ing them twice in LB, the selection procedure was repeated another 3 times. The cultures were then replica-plated on LB plates with and without tetracycline (10 µg/ml) and tetracycline-sensitive colonies were isolated (about 2-3%). The de-sired deletion mutants were distinguished from wild-type revertants and verified by sequencing of a 1.7 kb DNA fragment amplified from genomic DNA by PCR using primers 5'-attactcgaccagttccg and 5'-ttacgcaatggtatgatgc which are located outside the fragment cloned into the suicide vector pME3087 at position 2113 and 24385 (GenBank:NC_007005.1), respectively.

Construction of pOEAC

The sylA ORF was amplified from genomic DNA using the primers P7 (5'-cca-tcgatggagtagagtgatggc; ClaI site in italics, translation initiation codon indicated in bold) and P8 (5'-ggaattcttacaaaattcccatcttg; reverse primer; EcoRI site in ital-ics, reverse stop codon in bold), digested with ClaI and EcoRI, and cloned into the respective polylinker sites of the pME6001 (Gm^R) expression vector [34], thereby placing it under the control of the lacZ promoter. The resulting plasmid was named pOEA. As it turned out that pOEA did not confer gentamycin resistance in SRM$_{AF}$ medium, pOEAC was used, a derivative of pME6014 (Tet^R) [35], which, in addition to the lacZ::sylA chimeric gene, contained a sylC::lacZ reporter fusion gene in opposite orientation (the reporter gene is of no relevance in the present context). To construct pOEAC, the lacZ::sylA fusion gene was amplified from pOEA with primers P8 (see above) and P9 (5'-accgtccaacattaatgcagctgg; upstream of lac promoter; bold base complementary to position 987 of pBluescript vector (GenBank:X52329)) and joined with a sylC promoter fragment (position 5409-5649 of GenBank:AJ548826) that was amplified with primers P10 (5'-ctgcattaatgttggacggtctgc; bold base at position 5409) and P11 (5'-aactgcagtcat-gacggcctcggat; PstI site in italics, bold base at position 5649) by overlap extension PCR using primers P8 and P11. The resulting fragment was digested with EcoRI and PstI and cloned between the respective sites in the polylinker of pME6014.

Bacterial Infiltration of Rice Leaves and RNA Gel Blot Analysis

Bacterial strains were grown on a rotary shaker (220 rpm) over night at 28°C in LB containing, where appropriate, 10 µg/ml tetracycline. Bacteria were pelleted by centrifugation, washed twice in distilled water, resuspended in distilled water at an optical density at 600 nm (OD600) of 0.4 (approximately 10^8 cfu), and infiltrated into first leaves of 14-day-old rice plants (Oryza sativa cv. Loto; supplied by Terreni alla Maggia, Ascona, Switzerland) as described previously [23]. RNA was extracted from infiltrated leaves 16 h after infiltration and subjected to gel blot analysis using a ^{32}P-labeled Pir7b cDNA probe (GenBank:Z34270[23]) according to standard procedures [29].

HPLC Analysis and Mass Spectrometry of Syringolin A

To analyze conditioned media with regard to syringolin A content, Pseudomonas strains were grown in SRMAF medium [36,37] at 28°C for 60 h on a rotary shaker (220 rpm). Bacteria were pelleted by centrifugation and the supernatant was sterile filtered (0.22 µm pore size). Two-hundred-microliter aliquots were acidified with trifluoroacetic acid (TFA; 0.3% final concentration) and subjected to reverse-phase HPLC with a Reprosil 100-5 C$_{18}$ 250/4.6 column (Dr. Maisch GmbH, Ammerbuch-Entringen, Germany) on a Dionex UltiMate 3000 system (Dionex Corporation, Sunnyvale, CA). Elution was performed isocratically with 20% acetonitrile and 0.06% TFA in water at a flow rate of 1 ml/min.

High-resolution electrospray mass spectra were recorded on a Bruker maXis QTOF-MS instrument (Bruker Daltonics GmbH, Bremen, Germany). The samples were dissolved in MeOH and analyzed via continuous flow injection at 3 µl/min. The mass spectrometer was operated in positive ion mode with a capillary voltage of 4 kV, an endplate offset of -500 V, nebulizer pressure of 5.8 psig, and a drying gas flow rate of 4 l/min at 180°C. The instrument was calibrated with a Fluka electrospray calibration solution (Sigma-Aldrich, Buchs, Switzerland) that was 100 times diluted with acetonitrile. The resolution was optimized at 30'000 FWHM in the active focus mode. The accuracy was better than 2 ppm in a mass range between m/z 118 and 2721. All solvents used were purchased in best LC-MS qualities.

^{13}C-labeling and NMR Spectroscopy

Pss B301D-R was transformed with pOEAC and grown in LB containing 10 µg/ml tetracycline on a shaker at 28°C until an OD_{600} of approximately 0.5 was reached. Bacteria were collected by centrifugation, washed twice with SRM_{AF} medium, and taken up in SRMAF medium at an OD_{600} of 0.3. Fifty-ml cultures were inoculated with 0.01 volume of the bacterial suspension and incubated at 28°C on a shaker (220 rpm). After 48 h, $NaH^{13}CO_3$ (98%; Sigma-Aldrich, Buchs, Switzerland) was added to a final concentration of 70 mM and incubation was continued for 20 h. Bacteria were pelleted and syringolin A was isolated from sterile-filtrated conditioned media as described [4].

1H broadband decoupled 13C NMR spectra were recorded at 25°C on a Bruker Avance III 600 MHz spectrometer equipped with a cryogenic 5 mm CP-DCH probe head optimized for 13C detection. Two samples were prepared by dissolving 200 µg of labeled syringolin A in 130 µl DMSO-d6 and 5 mg of unlabeled syringolin A in 750 µl DMSO-d6, respectively. The labeled sample was transferred to a 3 mm Shigemi tube, the unlabeled sample was transferred to a regular 5 mm NMR tube. The spectral width in both spectra was 248.5 ppm, the transmitter was set to 100 ppm. The excitation pulse angle was set to 45°. The acquisition time was 2.1 s with a waiting time of 0.3 s between two scans. Both spectra were 1H broadband decoupled using the waltz16 composite-pulse decoupling scheme. The resulting fid consisted of 157890 total data points. For the unlabeled syringolin A sample 4000 scans were accumulated. For the labeled syringolin A sample 29605 scans were accumulated. Both spectra were zero filled to 131072 complex data points and processed using an exponential line broadening of 2 Hz. The samples contained no internal chemical shift reference and the spectra were referenced to the solvent peak (39.5 ppm). By comparison with chemical shifts listed in [1] the signals at 157.8 ppm and 132.8 ppm were assigned to the ureido CO group and the olefinic C at position 4 in the 3,4-dehydrolysine moiety, respectively.

Authors' Contributions

CR carried out the majority of experiments. The pPL3syl cosmid and the Δsyl deletion mutant were constructed by MT and MM, respectively. Mass spectrometry and NMR spectroscopy were performed and analyzed by LB and MOE, respectively. BS performed RNA gel blot analyses in the rice infiltration experiments. RD, CR, and BS designed experiments and RD wrote a draft manuscript. All authors provided critical inputs to the manuscript.

Acknowledgements

We thank Zsuzsa Hasenkamp for expert technical assistance, and Enrico Martinoia, Stefan Hörtensteiner, Markus Kaiser, and André Bachmann for discussions and helpful comments on the manuscript. Financial support by the Swiss National Science Foundation (grant 3100A0-115970 to RD) is acknowledged.

References

1. Wäspi U, Blanc D, Winkler T, Ruedi P, Dudler R: Syringolin, a novel peptide elicitor from Pseudomonas syringae pv. syringae that induces resistance to Pyricularia oryzae in rice. Mol Plant-Microbe Interact 1998, 11(8):727–733.

2. Wäspi U, Hassa P, Staempfli A, Molleyres L-P, Winkler T, Dudler R: Identification and structure of a family of syringolin variants: Unusual cyclic peptides from Pseudomonas syringae pv. syringae that elicit defense responses in rice. Microbiol Res 1999, 154:1–5.

3. Michel K, Abderhalden O, Bruggmann R, Dudler R: Transcriptional changes in powdery mildew infected wheat and Arabidopsis leaves undergoing syringolin-triggered hypersensitive cell death at infection sites. Plant Mol Biol 2006, 62:561–578.

4. Wäspi U, Schweizer P, Dudler R: Syringolin reprograms wheat to undergo hypersensitive cell death in a compatible interaction with powdery mildew. Plant Cell 2001, 13(1):153–161.

5. Groll M, Schellenberg B, Bachmann AS, Archer CR, Huber R, Powell TK, Lindow S, Kaiser M, Dudler R: A plant pathogen virulence factor inhibits

the eukaryotic proteasome by a novel mechanism. Nature 2008, 452(7188): 755–758.

6. Clerc J, Groll M, Illich DJ, Bachmann AS, Huber R, Schellenberg B, Dudler R, Kaiser M: Synthetic and structural studies on syringolin A and B reveal critical determinants of selectivity and potency of proteasome inhibition. Proc Natl Acad Sci USA 2009, 106(16):6507–6512.

7. von Döhren H, Dieckmann R, Pavela-Vrancic M: The nonribosomal code. Chem Biol 1999, 6(10):R273–R279.

8. Marahiel MA, Stachelhaus T, Mootz HD: Modular peptide synthetases involved in nonribosomal peptide synthesis. Chem Rev 1997, 97(7):2651–2673.

9. Finking R, Marahiel MA: Biosynthesis of nonribosomal peptides. Annu Rev Microbiol 2004, 58:453–488.

10. Amrein H, Makart S, Granado J, Shakya R, Schneider-Pokorny J, Dudler R: Functional analysis of genes involved in the synthesis of syringolin A by Pseudomonas syringae pv. syringae B301D-R. Mol Plant-Microbe Interact 2004, 17(1):90–97.

11. Fischbach MA, Walsh CT: Assembly-line enzymology for polyketide and non-ribosomal peptide antibiotics: Logic, machinery, and mechanisms. Chem Rev 2006, 106(8):3468–3496.

12. Hopwood DA: Genetic contributions to understanding polyketide synthases. Chem Rev 1997, 97(7):2465–2497.

13. Harada K, Fujii K, Shimada T, Suzuki M, Sano H, Adachi K, Carmichael WW: Two cyclic peptides, anabaenopeptins, a third group of bioactive compounds from the cyanobacterium Anabaena flos-aquae NRC–525–17. Tetrahedron Lett 1995, 36(9):1511–1514.

14. Murakami M, Shin HJ, Matsuda H, Ishida K, Yamaguchi K: A cyclic peptide, anabaenopeptin B, from the cyanobacterium Oscillatoria agardhii. Phytochemistry 1997, 44(3):449–452.

15. Gesner-Apter S, Carmeli S: Three novel metabolites from a bloom of the cyanobacterium Microcystis sp. Tetrahedron 2008, 64(28):6628–6634.

16. Okumura HS, Philmus B, Portmann C, Hemscheidt TK: Homotyrosine-containing cyanopeptolins 880 and 960 and anabaenopeptins 908 and 915 from Planktothrix agardhii CYA 126/8. J Nat Prod 2009, 72(1):172–176.

17. Williams DE, Craig M, Holmes CFB, Andersen RJ: Ferintoic acids A and B, new cyclic hexapeptides from the freshwater cyanobacterium Microcystis aeruginosa. J Nat Prod 1996, 59(6):570–575.

18. Matthew S, Ross C, Paul VJ, Luesch H: Pompanopeptins A and B, new cyclic peptides from the marine cyanobacterium Lyngbya confervoides. Tetrahedron 2008, 64(18):4081–4089.

19. Schmidt EW, Harper MK, Faulkner DJ: Mozamides A and B, cyclic peptides from a theonellid sponge from Mozambique. J Nat Prod 1997, 60:779–782.

20. Muller D, Krick A, Kehraus S, Mehner C, Hart M, Kupper FC, Saxena K, Prinz H, Schwalbe H, Janning P, et al.: Brunsvicamides A-C: Sponge-related cyanobacterial peptides with Mycobacterium tuberculosis protein tyrosine phosphatase inhibitory activity. J Med Chem 2006, 49(16):4871–4878.

21. Staskawicz B, Dahlbeck D, Keen N, Napoli C: Molecular characterization of cloned avirulence genes from race-0 and race-1 of Pseudomonas syringae pv. glycinea. J Bacteriol 1987, 169(12):5789–5794.

22. Wäspi U, Misteli B, Hasslacher M, Jandrositz A, Kohlwein SD, Schwab H, Dudler R: The defense-related rice gene Pir7b encodes an "alpha/beta hydrolase fold" protein exhibiting esterase activity towards naphthol AS-esters. Eur J Biochem 1998, 254:32–37.

23. Reimmann C, Hofmann C, Mauch F, Dudler R: Characterization of a rice gene induced by Pseudomonas syringae pv. syringae: Requirement for the bacterial lemA gene function. Physiol Mol Plant Pathol 1995, 46(1):71–81.

24. Senior E, Bull AT, Slater JH: Enzyme evolution in a microbial community growing on herbicide Dalapon. Nature 1976, 263(5577):476–479.

25. Smith JA, Métraux JP: Pseudomonas syringae pathovar syringae induces systemic resistance to Pyricularia oryzae in rice. Physiol Mol Plant Pathol 1991, 39(6):451–461.

26. Kanehisa M, Araki M, Goto S, Hattori M, Hirakawa M, Itoh M, Katayama T, Kawashima S, Okuda S, Tokimatsu T, et al.: KEGG for linking genomes to life and the environment. Nucleic Acids Res 2008, 36:D480–D484.

27. Golemi D, Maveyraud L, Vakulenko S, Samama JP, Mobashery S: Critical involvement of a carbamylated lysine in catalytic function of class D beta-lactamases. Proc Natl Acad Sci USA 2001, 98(25):14280–14285.

28. Maveyraud L, Golemi D, Kotra LP, Tranier S, Vakulenko S, Mobashery S, Samama JP: Insights into class D beta-lactamases are revealed by the crystal structure of the OXA10 enzyme from Pseudomonas aeruginosa. Structure 2000, 8(12):1289–1298.

29. Ausubel FM, Brent R, Kingston RE, Moore DD, Smith JA, Seidman JG, Struhl K: Current protocols in molecular biology. New York: Wiley and Sons; 1987.

30. Finan TM, Kunkel B, Devos GF, Signer ER: 2nd symbiotic megaplasmid in Rhizobium meliloti carrying exopolysaccharide and thiamin synthesis genes. J Bacteriol 1986, 167(1):66–72.

31. Christensen BB, Sternberg C, Andersen JB, Palmer RJ, Nielsen AT, Givskov M, Molin S: Molecular tools for the study of biofilm physiology. In Methods Enzymol. San Diego: Academic Press; 1999:20–42.

32. Voisard C, Bull CT, Keel C, Laville J, Maurhofer M, Schnider U, Défago G, Haas D: Biocontrol of root diseases by Pseudomonas fluorescens CHA0: current concepts and experimental approaches. In Molecular ecology of rhizosphere microorganisms. Edited by: F. OG, D. D, B B. Weinheim, Germany: VCH Publishers; 1994:67–89.

33. Simon R, Priefer U, Puhler A: A broad host range mobilization system for in vivo geneic engineering: transposon mutagenesis in Gram-negative bacteria. Bio-Technology 1983, 1(9):784–791.

34. Blumer C, Heeb S, Pessi G, Haas D: Global GacA-steered control of cyanide and exoprotease production in Pseudomonas fluorescens involves specific ribosome binding sites. Proc Natl Acad Sci USA 1999, 96(24):14073–14078.

35. Schnider-Keel U, Seematter A, Maurhofer M, Blumer C, Duffy B, Gigot-Bonnefoy C, Reimmann C, Notz R, Defago G, Haas D, et al.: Autoinduction of 2,4-diacetylphloroglucinol biosynthesis in the biocontrol agent Pseudomonas fluorescens CHA0 and repression by the bacterial metabolites salicylate and pyoluteorin. J Bacteriol 2000, 182(5):1215–1225.

36. Gross DC: Regulation of syringomycin synthesis in Pseudomonas syringae pv. syringae and defined conditions for its production. J Appl Bacteriol 1985, 58:167–174.

37. Mo Y-Y, Gross DC: Plant signal molecules activate the syrB gene, which is required for syringomycin production by Pseudomonas syringae pv. syringae. J Bacteriol 1991, 173:5784–5792.

Mapping of Protein Phosphatase-6 Association with its SAPS Domain Regulatory Subunit Using a Model of Helical Repeats

Julien Guergnon, Urszula Derewenda, Jessica R. Edelson and
David L. Brautigan

ABSTRACT

Background

Helical repeat motifs are common among regulatory subunits for type-1 and type-2A protein Ser/Thr phosphatases. Yeast Sit4 is a distinctive type-2A phosphatase that has dedicated regulatory subunits named Sit4-Associated Proteins (SAPS). These subunits are conserved, and three human SAPS-

related proteins are known to associate with PP6 phosphatase, the Sit4 human homologue.

Results

Here we show that endogenous SAPS subunit PP6R3 co-precipitates half of PP6 in cell extracts, and the SAPS region of PP6R3 is sufficient for binding PP6. The SAPS domain of recombinant GST-PP6R3 is relatively resistant to trypsin despite having many K and R residues, and the purified SAPS domain (residues 1-513) has a circular dichroic spectrum indicative of mostly alpha helical structure. We used sequence alignments and 3D-jury methods to develop alternative models for the SAPS domain, based on available structures of other helical repeat proteins. The models were used to select sites for charge-reversal substitutions in the SAPS domain of PP6R3 that were tested by co-precipitation of endogenous PP6c with FLAG-tagged PP6R3 from mammalian cells. Mutations that reduced binding with PP6 suggest that SAPS adopts a helical repeat similar to the structure of p115 golgin, but distinct from the PP2A-A subunit. These mutations did not cause perturbations in overall PP6R3 conformation, evidenced by no change in kinetics or preferential cleavage by chymotrypsin.

Conclusion

The conserved SAPS domain in PP6R3 forms helical repeats similar to those in golgin p115 and negatively charged residues in interhelical loops are used to associate specifically with PP6. The results advance understanding of how distinctive helical repeat subunits uniquely distribute and differentially regulate closely related Ser/Thr phosphatases.

Background

Helical repeat motifs such as ANK, HEAT, and ARM are thought to primarily mediate protein-protein interactions (see reviews[1-3]). Helical repeat motifs are a recurrent theme among regulatory subunits for different protein Ser/Thr phosphatases. Best studied is the A or PR65 subunit of PP2A, an all-helical subunit first designated to consist of Armadillo (ARM) sequence repeats, that were later called HEAT repeats [4], a name derived from proteins with related sequence motifs: Huntingtin's, elongation factor, A subunit of PP2A and TOR. The 3D structure of the A subunit of PP2A alone [5], as a dimer bound to the PP2A catalytic subunit [6], and as a scaffold to assemble PP2A heterotrimers [7-9], showed the all-helical organization and revealed differences in overall conformation due to association with the other subunits. The extended arc of helices is shaped like a banana in the monomer or heterodimer and closes to a horseshoe-shaped conformation in the heterotrimer. In addition, in the ABC trimers the regulatory B'56 subunit of PP2A was found to be a HEAT-like helical repeat protein that contacts both the A and C subunits. The structure of B'56 was unexpected because it was not predicted based on sequence alignments with other HEAT-repeat proteins. Another example of helical repeat motifs in protein phosphatase subunits is the MYPT1 subunit for PP1, with 8 ankyrin repeats [10]. In the 3D structure these repeats form an arc of alpha helices to engage the top surface of the PP1 catalytic subunit and enwrap the C-terminal tail that protrudes from the top surface of the subunit. Both the ANK repeats as well as a separate structural element consisting of an alpha helix plus a neighboring strand with the canonical RVxF motif make contacts with the PP1 catalytic subunit. Based on these examples there is the expectation that other phosphatase regulatory subunits might be comprised of helical repeat structures and use these repeats to mediate subunit-subunit association.

The yeast Sit4 phosphatase is related in sequence and properties to members of the type-2A family of protein Ser/Thr phosphatases [11]. Strains with temperature-sensitive mutations (sit4ts) are rescued by ectopic expression of human PP6 [12], but not the close relative PP4, showing functional complementation across species, but specificity for the individual type of catalytic subunit. The results argue for distinct lines of evolutionary descent for PP2A, PP6 and PP4, with a high degree of conservation within each line. Yeast Sit4 has multiple associated subunits that co-immunoprecipitate, first named Sit4-Associated Proteins (SAP) [13]. Sequence alignments using a region common in the yeast SAP identified SAPS in various species, including three human proteins (KIAA1115, KIAA0685 and C11orf23), which were renamed PP6R1, PP6R2 and PP6R3 and shown to co-precipitate with PP6, but neither PP4 nor PP2A [14]. The sequence motif in yeast and human proteins, as well as in other species, has been designated as a "SAPS" domain by PFAM http://pfam.sanger.ac.uk/ webcite. These SAPS domain proteins are proposed to function as specific regulatory subunits for PP6. Truncation of the C-terminal region of PP6R1 did not compromise co-precipitation with PP6, showing that the designated SAPS domain was sufficient for binding the catalytic subunit.

The physiological function(s) of this family of SAPS domain proteins as specific regulatory subunits is not yet well understood. Knockdown of individual SAPS domain subunits by siRNA mimics knockdown of PP6 catalytic subunit itself in terms of effects on putative individual substrates such as IκBε and DNA-PK [14,15]. The results argue that all three SAPS subunits can individually associate with PP6, but the different PP6 complexes have non-overlapping substrate specificity. The conservation of SAPS function in PP6R1, PP6R2 and PP6R3 subunits is demonstrated by their ability to co-precipitate endogenous Sit4 when expressed in yeast strains deleted for their endogenous SAPs [16]. Furthermore, the human SAPS proteins partially restored functions in these strains. Proteomics discovered association of PP6R1 with multiple ankyrin repeat domain proteins (Ankrd), suggesting a heterotrimeric organization of PP6 [17]. This makes the mammalian PP6 different from the yeast Sit4 because the Ankrd genes are not found in yeast and there is no evidence for Sit4 heterotrimers. This opens the possibility that the Ankrd subunits function to target and regulate PP6 actions in addition to the SAPS subunits. Here we analyzed the properties of the PP6R3 subunit using partial proteolysis, deletions and co-precipitation assays from cells. The SAPS domain sequences were used to produce alternate structural models of helical repeat motifs, and these models were used to select sites for point mutations that were assayed by co-precipitation. The results indicate the SAPS domain subunits of PP6 use a newly-found variation of the helical repeat theme to achieve selective recognition of PP6c.

Methods

Antibodies, Immunoprecipitation and Immunoblotting

Anti-FLAG antibodies were purchased from Sigma-Aldrich and used at a dilution of 1:5000 (Rabbit) or 1:3000 (mouse). Goat anti-rabbit Alexa Fluor 680 and donkey anti-sheep Alexa Fluor 680 were purchased from Molecular Probes and Invitrogen and used at a 1:5000 dilution. Goat anti-mouse IRDye 800 and anti-chicken IRDye 800 antibodies were purchased from Rockland Immunochemicals and used at a 1:5000 dilution. Rabbits were immunized with purified recombinant GST made in bacteria, and anti-GST antibodies were affinity-purified from serum and used at a 1:5000 dilution. Antibody against PP6 [14] was used at a 1:3000 dilution. Sheep were immunized with purified recombinant GST-PP6R3 made in bacteria, and antibodies were affinity-purified from serum against MBP-PP6R3 and used at a 1:2000 dilution. Anti-PP6R3 beads were prepared as follows: 200 µl of protein-G agarose beads were incubated with 400 µg of affinity purified anti-PP6R3 in 2 ml at room temperature for 1 h. Beads were then washed twice by centrifugation with 1 ml of 0.2 M sodium borate solution, pH 9.0. Beads were resuspended in 1 ml of sodium borate and a final concentration of 20 mM disuccinimdyl pimelimate added. After incubation at room temperature for 30 min the coupling reaction was stopped by washing the beads once by centrifugation, with resuspension and incubation in 0.2 M ethanolamine for 2 h at room temperature. Beads were washed twice by centrifugation and resuspended in PBS. Chicken anti-PP6R3 IgY antibodies were prepared using GST-PP6R3 as immunogen by Aves Laboratories (California).

For immunoprecipitation cells were washed with ice-cold PBS then lysed with 20 mM Tris-HCl, pH 7.5, 150 mM NaCl, 1% NP-40, 5 mM EDTA, 5 mM NaF and protease inhibitor mixture set V, EDTA-free (Calbiochem). After 15 min on ice cells were scraped and the suspension transferred to a tube, mixed on a vortex, and centrifuged at 20,000 × g for 20 min. Supernatants were used for immunoprecipitation with 30 μl of immobilized anti-PP6R3 or anti-FLAG beads (M2, Sigma) overnight at 4°C with gentle rotation. Beads were washed 3 times with the lysis buffer then boiled 5 min in 2 × SDS sample buffer. SDS-PAGE was done using acrylamide/Bis 29:1 as previously described [14] or using pre-cast CRITERION gradient gels (BioRad). Gels were stained with Coomassie blue or GelCode (Thermo Scientific), or proteins were transferred by electrophoresis onto nitrocellulose for immunoblotting, and developed using LI-COR Odyssey Infrared Imaging System (LI-COR biotechnology). This scanner provides quantitative analysis with a extended range of linear response.

Plasmids and PCR

DNA of human PP6R3 was amplified by PCR from HeLa cDNA generated by Thermoscript poly(dT) reverse transcription-PCR (Invitrogen) following the manufacturer's protocol. The primers for PCR of PP6R3 were 5'-GAA TTC ATG TTT TGG AAA TTT GAT CTT C-3' as the forward primer and 5'-CTC GAG CAC TTC AGT GAA TGG CCC TGT ATC ACT G-3' as the reverse primer. PP6R3 fragments 1-355 and 1-513 were generated using the same forward primer as PP6R3 full length, and 5'-CTCGAGTCAAAGCAG-GCTGGATATCAACCTAATG-3' and 5'-CTCGAGTCAGTTCCTCTTGT-TAGTTTCTCCTAAG-3' as reverse primers respectively. PP6R3 512-873 was generated using 5'-GAATTCACGGTAGATCTAATGCAAC-3' as the forward primer and 5'-CTCGAGTCATACAGGGCCATTCACTGAAGTG-3' as the reverse primer. For directed mutagenesis, the QuickChange site-directed mutagenesis kit was used according to manufacturer's protocol. Primers for E63-E64 to K were 5'-GTCTCATTCATTATAAAAAAACCACCTCAAGACATGGATG-3' as the forward primer and 5'-CATCCATGACTTGAGGTGGTTTTTTTATAAT-GAATGAGAC-3' as the reverse primer. Primers for mutation of D113 to R were 5'-GCTTCCTCCTAAACCGTTCCCCTTTGAATCCACTAC-3' and 5'-GTAGTGGATTCAAAGGGGAACGGTTTAGGAGGAAGC-3' as forward and reverse primers respectively. Primers to mutate E204-E205 to K were 5'-GTTCATCCATCGCAAAAAAAGATCGACATTCAAATGC-3' as the forward primer and 5'-GCATTTGAATGTCGATCTTTTTTTTGCGATGGAT-GAAC-3' as the reverse primer. Primers for mutation of E259-E262 to K were 5'-CAAATATTTTCCACAAGAAGAAAAATAAGTCAGCCATAGTCAG-3' and   5'-CTGACTATGGCTGACTTATTTTTCTTCTTGTGGAAAAT-ATTTG-3' as forward and reverse primers respectively.

Recombinant Protein Production and Analyses

Full length PP6R3 (accession Q5H9R7) was subcloned in pGEX-4T-1 vector for production of recombinant GST-PP6R3 in BL-21 strain of E. coli that was purified using glutathione Sepharose following manufacturer's instructions. PP6R3 SAPS domain (residues 1-513) was subcloned by PCR and ligated into pET30 vector, expressed in BL21 cells, and purified by metal ion affinity chromatography using Ni-NTA Agarose (Qiagen).

Circular dichroic (CD) spectrum of purified recombinant PP6R3(1-513) at 0.2 mg/ml in Tris HCl pH 7.4 with 0.15 M NaCl was recorded with a Aviv model 215 spectropolarimeter.

Purified GST-PP6R3 (2 mcg) was incubated with 10 ng of TPCK-trypsin in 20 mM Tris pH 7.5 at room temperature for times ranging from 0 to 10 minutes. Reaction was stopped by transferring aliquots of the reaction into ice-cold denaturing buffer (final concentrations 2% SDS, 50 mM Tris pH 7.5, in 10% glycerol). Samples were analyzed by SDS-PAGE and immunoblotting. Molecular size standards were BioRad Precision Plus.

Cell Culture and Transfections

HEK293, HEK293T and HepG2 cells were grown in MEM medium with 2 mM L-glutamine and 10% FBS. HeLa cells were grown in DMEM medium containing 10% FBS. Jurkat cells were grown in RPMI medium with 2 mM L-glutamine and 10% FBS. Cells were grown at 37°C in humidified 5% CO_2 atmosphere. HEK293 cells at 40-50% confluence in 10 cm dishes were transiently transfected using 4-10 µg of plasmid DNA mixed with Arrest-in (Open Biosystems) for 20 min at room temperature before addition to cells. Cells were harvested after 24 h and extracts prepared for immunoprecipitation.

Sequence Alignments and Structure Modelling

Amino acid sequences of PP6R3 were not recognised as helical repeat motifs by standard algorithms (including by PFAM, which was used to define the SAPS family). The PP6R3 amino acid sequence did not align with signature sequence motifs for either the ARM or HEAT helical repeats. Furthermore, the human PP6R3 sequence did not align with any other protein of known 3D structure, when using standard algorithms. We used the ALSCRIPT program [18] to align sequences as shown in Figure four.

Modelling of the SAPS protein was carried out using program REP [17]. REP was designed to identify structural repeats from protein sequences. The program has been trained on profiles from several known ARM and HEAT repeat proteins, but did not identify repeat motifs in the PP6R3 sequence when used with standard thresholds. REP in a low-confidence, no-thresholds-applied mode detected sequence repeats in PP6R3, but failed to assign them as ARM, HEAT or ANK repeats.

As an alternative we opted to use the 3D-Jury protein structure prediction Meta Server http://meta.bioinfo.pl/ webcite. The metaserver first generates 3D models using diverse structure prediction methods, and then compares and scores models, based on their consistency [19]. This approach allowed us to generate possible models for the SAPS domain, based on other proteins that contain either ARM or HEAT repeats. These models were depicted as structures in Figure five using PyMOL http://pymol.sourceforge.net/ webcite.

Chymotrypsin Digestion of Wild-Type vs. Mutated FLAG-PP6R3

Human 293T cells at <40% confluence in 100 mm dishes were transfected using 30 µl Arrest-In (Open Biosystems) diluted in 50 µl Opti-MEM with 5 µg of plasmid DNA encoding either wild type PP6R3 or quadruple mutant PP6R3 (E204K, E205K, E259K, E262K). After 24 h, cells were lysed using 1% NP-40 buffer and centrifuged at 13,000 rpm for 15 min. Aliquots of 500 µl of the supernatant were incubated with 50 µl M2 Agarose beads (Sigma, A2220) overnight at 4°C with rocking. Beads were washed 3× with 100 µl of wash buffer A (1% NP-40 plus 2 mM ATP and 5 mM $MgCl_2$), two times with 100 µl wash buffer B (50 mM Tris pH 8.0, 500 mM NaCl), and one time with 100 µl wash buffer C (50 mM Tris pH 8.0, 150 mM NaCl). The beads were resuspended in 120 µl wash buffer C and 20 µl removed as time zero control. Chymotrypsin was added to a final concentration of 0.5 ng/µl. At various times 20 µl aliquots were removed, mixed with 20 µl 2 × SDS buffer and heated to 100°C for 5 min.

Samples were resolved by 4-15% gradient SDS-PAGE, transferred onto nitrocellulose by a semi-dry protocol, and the filter blocked with 5% non-fat milk in Tris-buffered saline plus 1% Tween 20. The filters were probed with chicken anti-R3 antibodies (1:2000) or anti-FLAG antibodies (Sigma F7425, 1:1000), developed with fluorescent secondary antibodies, and scanned 2D images were captured by an Odyssey 2D infrared scanner (LiCor Industries). As a control a duplicate sample of washed beads resuspended in 120 µl wash buffer C was incubated at 100°C for 5 min. Following centrifugation the supernatant with denatured GST-PP6R3 was incubated with chymotrypsin, and samples at various time points processed as described above. Molecular size standards were BioRad Precision Plus.

Results

The SAPS domain appears in three human proteins, based on a common region of sequence identity and similarity (14). We prepared individual antibodies to each and found that PP6R3 (a.k.a. SAPS3) was most efficiently immunoprecipitated from HEK293 cell extracts (Figure 1). PP6R3 co-precipitated endogenous PP6 catalytic subunit (PP6c) that was detected by immunoblotting (Figure 1A, lane 3). Immunoprecipitation was specific because neither PP6c nor PP6R3 was in control samples prepared using blank beads or non-immune primary antibody (Figure 1A, lanes 1, 2). After two rounds of immunoprecipitation the extracts were fully depleted of PP6R3 (Figure 1B, lane 2) as well as 50% of the endog-

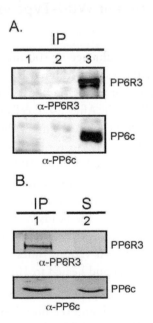

Figure 1. Co-immunoprecipitation of endogenous PP6R3 and PP6c. (A) Immunoprecipitates (IP) were prepared from HeLa cells with pre-immune sheep serum (lane 1), blank beads (lane 2) or anti-PP6R3 sheep antibody (lane 3). Precipitates were analyzed by immunoblotting for PP6R3 (upper panel) and PP6c (lower panel). (B) Co-immunoprecipitation of PP6R3 and PP6c from HepG2 cells with anti-PP6R3 antibody (lane 1). The PP6R3 and PP6c remaining unbound were detected by immunoblotting the supernatant (S, lane 2) for PP6R3 (upper panel) and PP6c (lower panel)

enous PP6c. The other 50% of PP6c in the supernatant presumably was associated with other subunits besides PP6R3. The same results were obtained with co-precipitation from extracts of HeLa, Jurkat and HepG2 cells (not shown). We concluded that endogenous PP6R3 stably associates with about half of the total PP6c in tissue culture cell lines, making it the major partner for PP6 and a suitable candidate for further study as representative of the SAPS domain subunits.

Trypsin Digestion and CD Spectrum of Recombinant PP6R3

We expressed recombinant GST-PP6R3 fusion protein in bacteria, affinity-purified the protein and subjected it to partial proteolysis by trypsin to examine its fragmentation (Figure 2A). The purified fusion protein migrated as a single band at 140 kDa and, as expected, was reactive with both anti-GST and anti-PP6R3 antibodies. The actual molecular mass of PP6R3 is 97.7 kDa (873 residues), so the fusion protein migrates much less (at a higher Mr) than predicted. Another protein at ~80 kDa was detected by anti-GST immunoblotting, but not by anti-PP6R3 blotting, and this protein was not digested by trypsin under the conditions used, therefore we used it as a convenient loading control for aliquots taken at different time points. Multiple trials were conducted to optimize conditions for time-dependent conversion of the GST-PP6R3 into fragments. Digestion with trypsin at 1/5000 (w/w) at room temperature and pH 7.5 (shown in Figure 2) gave progressive loss of the full length GST-PP6R3 fusion protein that was nearly complete by 10 min. Treatment of GST with trypsin under identical conditions showed that GST was relatively resistant to digestion (Figure 2A, left 2 lanes). Therefore, we could use anti-GST immunoblotting to track cleavage of GST-PP6R3. The full-length fusion protein was converted into four major fragments that were recognised by anti-GST (Figure 2A) as well as with anti-PP6R3 (not shown). One fragment formed first, as early as 30 sec, and remained as the primary product after 10 min. All the major products were >60 kDa, suggesting that trypsin cleaved following the GST fused to the N terminus at a point ca. 330-350 residues into the PP6R3 sequence. We concluded that the N-terminal region of PP6R3 up to that point has a conformation in solution that limits digestion by trypsin, even though there are many lysine and arginine residues in this region.

In addition, a recombinant His6-tagged PP6R3 protein, residues 1-513, was purified from bacteria. The CD spectrum was recorded (Figure 2B) and the purity of the protein verified by Coomassie staining after SDS-PAGE (Figure 2C). The CD spectrum with minima at 208 and 222 nm is characteristic of proteins that are predominantly made of alpha helices [20]. Together these results suggest that the SAPS domain is composed of alpha helical secondary structures in a conformation that limits trypsin digestion.

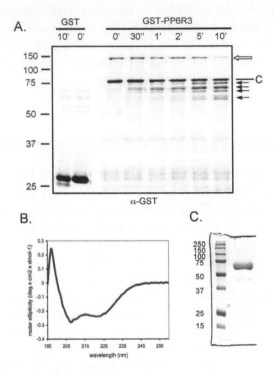

Figure 2. Trypsin fragmentation of GST-PP6R3 and CD spectrum of SAPS domain. (A) Recombinant GST-PP6R3 (upper band, open arrow) was digested with trypsin from 0 to 10 min. The GST-PP6R3 and fragments were detected using an anti-GST antibody and a non-specific band (labelled C) that was not digested served as loading control Trypsin caused time-dependent formation of four digestion products (arrows). GST was digested under identical conditions for 10 min (left 2 lanes). The migration of molecular size standards is indicated to the left side of the frame. (B) Circular dichroic (CD) spectrum of purified recombinant PP6R3(1-513) plotted as molar ellipticity (x 10-5) vs. wavelength in nm. (C) Recombinant His6 tagged SAPS domain residues 1-513, purified and stained with Coomassie after SDS-PAGE. Left lane, size standards in kDa; right lane is protein used for CD spectrum.

Mapping PP6 Binding to SAPS Domain in PP6R3

To map the region of PP6R3 required for PP6c binding we expressed the following in HEK293 cells: 1) FLAG-tagged PP6R3 (full-length, residues 1-873), 2) a SAPS region (residues 1-513), 3) a C-terminal region (512-873) and 4) a truncated SAPS region (1-355) corresponding approximately to the trypsin-resistant N terminus. Cell extracts and FLAG immunoprecipitates were prepared and immunoblotted for FLAG and endogenous PP6c (Figure 3). Full length PP6R3 effectively co-precipitated PP6c (lane 1), as did the SAPS region (residues 1-513; lane 3). Less PP6c was recovered with 1-513 compared to full-length, but we attributed this to lower expression level of the 1-513 protein. The FLAG immunoprecipitate in lane 3 had about equal amounts of a protein corresponding to residues

1-513 and a smaller FLAG-tagged protein fragment. This smaller fragment is not expected to bind PP6c because of the lack of co-immunoprecipitation with the 1-355 protein of about the same size in the immunoblot (lane 2). We noted that endogenous proteolysis generated nearly the same-sized fragments (about 340 residues of PP6R3) in lanes 1, 2 and 3. PP6c did not co-precipitate at all with a C-terminal region of PP6R3 (residues 512-873) (Figure 3, lane 4). Previous data with a different SAPS subunit, PP6R1, showed residues 1-465 were sufficient for co-precipitation of PP6c, whereas residues 462-825 did not co-precipitate PP6c [14]. These new results show that in PP6R3 a region of residues 1-355 was not sufficient, but residues 1-513 could co-precipitate endogenous PP6c.

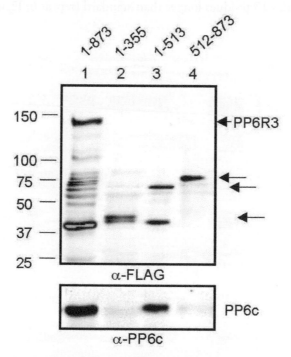

Figure 3. Deletion mapping of PP6c binding to PP6R3. FLAG-tagged full length PP6R3 (lane 1), or residues 1-355 (lane 2), 1-513 (lane 3) or 512-873 (lane 4) were expressed in HEK293 cells and immunoprecipitated with anti-FLAG antibody. FLAG-tagged proteins (upper panel) and co-immunoprecipitated endogenous PP6c (lower panel) were detected by immunoblotting.

Prediction of 3D Organization of the SAPS Domain

Alternative models for the PP6R3 SAPS region were generated by sequence alignments (Figure 4) and structural modeling by the 3D-Jury Meta Server, using proteins containing either ARM or HEAT repeats. The modelling fits regions of sequence into putative alpha helices and predicts surface loops between the helices. The highest score models (Figure 5) were based on: 1) importin alpha ARM repeat structures (PDB ID 2C1T, 1IAL, 1Q1S) (line 2, Figure 4), and 2) beta-catenin ARM domain, (1I7W, 1JDH) (line 3 Figure 4). Models based on HEAT repeats in beta importin (2BKU) and the PP2A PR65 subunit (1B3U) had lower scores, but were included as possible alternatives in Figure 4 (line 4 and 5). An unusual, new ARM repeat structure of a vesicular transport factor called golgin p115 was published recently [21]. This structure offers another alternative for modelling of PP6R3. In golgin p115, helical repeats 7-9 contain extended loops between helix 1 and 2, not previously seen in other ARM domains. This novel structure raises the possibility that residues ~223-276 in PP6R3 form a single elongated repeat, ~15 residues longer than standard (repeat 6, Figure 4).

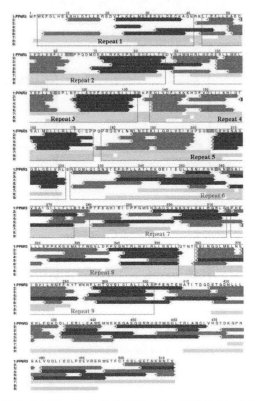

Figure 4. Alternative models for helical repeats in the SAPS domain of PP6R3. Line 1. PP6R3 primary sequence, residues 1-513. Positions of the mutations tested in Fig. 6 are highlighted in green. Lines 2-6. 3D-jury models based on: line 2- importin α; line 3- β-catenin; line 4- importin β; line 5- p115. Convex surface helices are red and concave helices are blue cylinders. Lines 7-9 are REP low confidence predictions of: line 7- HEAT; line 8- ARM; line 9- ANK repeats. For consistency with the models we have highlighted these areas pink for convex and light blue concave surfaces. Proposed helical repeat regions are boxed and labelled 1-9. We are less confident in repeats 6-9, labelled grey, compared to repeats 1-5 labeled in black.

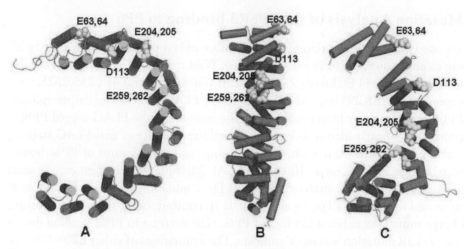

Figure 5. Structural models for SAPS domain in PP6R3. The sequence of PP6R3 residues 1-513 was used to produce alternate models based on known helical-repeat proteins (left to right): A- PP2A A subunit, B- beta-catenin, C- p115 golgin. Predicted alpha helices are shown in red (convex) and blue (concave), with intervening loops as strands. Side chains of E63, E64, D113, E204, E205, E259 and E262 are shown as space-filling models in yellow.

The structure of a HEAT repeat is similar to the structure of an ARM repeat, despite the fact that the ARM motifs consist of three helices, while HEAT motifs have only two. In both cases, the repeats are stacked together to form a super-helix (or solenoid) that in the case of PP2A PR65 and importin are curved (Figure 5A), while the structures of importin or beta-catenin are elongated, with less curvature (Figure 5B). The helices in the model based on the golgin p115 structure are less uniform and their arrangement is less compact (Figure 5C). There are predicted surface loops in the golgin p115-based structure that do not appear in the other models.

We expected that the SAPS predicted structure would resemble the HEAT helical repeats in the PP2A PR65 subunit, because, after all, this protein fold binds the PP2A catalytic subunit that is nearly identical to PP6. However, this model had the lowest 3D-Jury score. The models in both Figures 4 and 5 agree in assignment of the boundaries of the helical motifs in the N terminal residues 1-230, and therefore we were confident in the phasing of these intra-repeat loops. The assignments of helices and loops past residue 230 was more difficult and the models diverged in these segments. We speculated that the interactions between PP6R3 and the PP6 catalytic subunit could be mediated by the intra-motif loops and not the inter-motif loops. We used site-directed mutagenesis to generate charge reversal mutations in acidic residues that were predicted from the modelling to be in the loops. This was to test whether these sites were required for binding PP6c and therefore allow us to discriminate between the models in Figure 5.

Mutation Analysis of the PP6R3 Binding to PP6

We used transient expression of FLAG-PP6R3 and co-precipitation to assay binding to endogenous PP6c in cells (Figure 6). Wild type FLAG-PP6R3 and the following mutants 1) E63,64K; 2) D113K; 3) E204,205K and 4) E259,262K were expressed in HEK293 cells, and the amount of FLAG protein and co-precipitated PP6c determined by immunoblotting. The recovery of the FLAG-tagged PP6R3 proteins was nearly identical (Figure 6B) and the intensity of anti-FLAG staining was used to normalize the anti-PP6c staining, with the amount of PP6c bound to wild type PP6R3, set as 100 (Figure 6A). Multiple independent experiments gave results that were analyzed together. Dual mutation of E63 and E64 to K increased recovery of PP6c by about 30%. In contrast, other negative to positive charge mutations reduced binding of PP6c. The decrease in PP6c binding due to the D113R mutation was not significant. Dual mutations of either E204-E205 or E259-E262 produced statistically significant 25 to 35% reduction in PP6c binding. Combination of the dual mutations of 204/205 plus 259/262 resulted in the largest effect compared to wild type, an overall 80% reduction in co-precipitation of endogenous PP6c. These data indicated that the negatively charged residues in the predicted loops at residues 204, 205, 259 and 262 in PP6R3 were likely involved in association with PP6c.

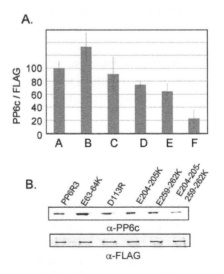

Figure 6. Mapping PP6c binding by charge-reversal mutations in PP6R3. (A) FLAG-tagged full length (FL) PP6R3 (A), and mutants E63-64K (B), D113R (C), E204-205K (D), E259-262K (E) and E204-205-259-262K (F) were expressed in HEK293 cells and immunoprecipitated using immobilized anti-FLAG antibody. Co-precipitated PP6c was quantified by fluorescent immunoblotting and normalized for the amount of FLAG-tagged protein. Results were replicated in 3 independent experiments and plotted as mean +/- SD. (B) Immunoblot of co-precipitated endogenous PP6c from one experiment (upper panel) and the FLAG-tagged PP6R3 proteins (lower panel).

Mutations Reduce PP6c Binding Without Change in Cleavage by Chymotrypsin

Mutations that result in a loss-of-function always leave open the possibility that alteration of overall protein conformation causes the loss of function, rather than a local change of a particular side chain. This is even true for charged resides that are likely to appear on the protein surface. To compare the conformation of wild type and quadruple mutated (204/205/259/262) PP6R3 we utilized partial proteolysis by chymotrypsin (Figure 7). This protease cleaves C terminal to aromatic residues and therefore mutation of charged residues should not affect its reaction. Indeed, we observed essentially identical patterns of chymotrypsin digestion for wild type and the quadruple mutant FLAG-PP6R3. Both the kinetics of the reaction over the first 10 min and the sizes of fragments formed were indistinguishable. The fragments were reactive with both anti-FLAG (upper panels) and anti-PP6R3 antibodies (lower panels), indicating that the primary sites of proteolysis were toward the C terminal region of the protein. We noted that the N terminal region (~37 kDa) of PP6R3 was relatively less susceptible to chymotrypsin, because a FLAG-tagged fragment of this size persisted longer than other fragments formed within the first min of digestion (lanes 3, 4, and 9, 10). These patterns of fragmentation reflected conformational constraints on proteolysis because denaturation of the FLAG-PP6R3 resulted in its complete degradation in less than one min in a parallel reaction (lanes 13-14). We concluded that the mutation of the negative to positive charges in the predicted loops that severely reduced binding of PP6c did not cause a major change in conformation of PP6R3.

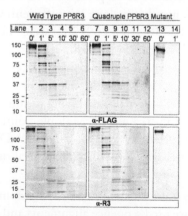

Figure 7. Conformational integrity of quadruple mutated FLAG-PP6R3. FLAG-tagged PP6R3 wild type (lanes 1-6) and the quadruple mutant E204K, E205K, E259K, E262K (lanes 7-14) were expressed in HEK293T cells, recovered by immunoprecipitation and digested with chymotrypsin for indicated periods of time. Aliquots of the reactions were resolved by SDS-PAGE and immunoblotted for both FLAG (upper panels) and PP6R3 (lower panels). As a control quadruple mutant protein eluted from beads (lane 13) was denatured and digested with the same amount of chymotrypsin for one min (lane 14) and analyzed by immunoblotting. Lines on left edge of panels show migration of molecular size standards.

Discussion

This study examined the properties of PP6R3, as an example of the conserved family of SAPS domain subunits that are specific for binding to PP6 phosphatase. The PP6R3 co-precipitated half the endogenous PP6c in different cell lines, showing it was the major binding partner for PP6, compared to the other SAPS subunits, and other proteins known to bind PP6c, such as alpha4 [22-24] and TIP [25], a.k.a. TAB4 [26]. Expression of the human SAPS subunits in yeast allowed for the co-precipitation of endogenous Sit4 with each of the SAPS. However, only PP6R3 fully rescued strains deleted for all yeast SAPS with restoration of budding, indicative of cell cycle progression [16]. We conclude that PP6R3 is a dominant and functionally conserved cellular partner for PP6c, making it a suitable representative of the family of SAPS proteins.

Our hypothesis is that the SAPS region of PP6R3 recognizes PP6c using alpha helical repeats that present residues in the inter-helical loops for subunit-subunit interaction. This follows in part from what is known about the PP2A complex with its scaffold A or PR65 subunit [6-8], but diverges in specific details. The N terminal 513 residues of PP6R3 are sufficient for stable association with PP6c, showing truncation of the C terminal 350+ residues of PP6R3 does not eliminate, or even diminish binding of PP6c. The 1-513 fragment of PP6R3 supports co-precipitation of PP6c, but a 1-355 fragment is not sufficient, suggesting that residues between 355 and 513 are critical, either for direct contact with the catalytic subunit, or for maintenance of the conformation of the SAPS domain. The inability of the 1-355 fragment to co-precipitate PP6c does not imply that it is not required or involved, just that it alone is insufficient. Clearly, point mutations within the 1-355 region (at 204, 205, 259, 262) were effective at nearly eliminating PP6c binding to full length PP6R3. This at least implies that these residues are somehow required and even dominant, though not alone sufficient, for stable subunit-subunit association. We speculate that there are contacts between PP6c and two regions of the SAPS domain, involving residues in the 200-265 region as well as residues in the 355-513 region. Multiple points of contact may well be necessary to achieve the specificity for PP6 vs. PP2A. Determination of the co-crystal structure will be needed to visualize the spatial organization of the SAPS domain and interactions with PP6c.

As a preliminary step in understanding the SAPS domain organization, and as a guide for mutagenesis we used sequence alignments and jury modelling methods to produce models of the SAPS region of PP6R3. The sequences of SAPS domains are sufficiently conserved so as to allow discovery of orthologs in species from yeast to mammals [14], and have been assigned a Pfam http://pfam.sanger.ac.uk/ designation, however the sequences do not allow a match to any structure based on available algorithms. The SAPS domain is predicted to have regions of alpha helical secondary structure, and the CD spectrum of an isolated SAPS domain indicates predominantly alpha-helical organization. We turned to jury modelling methods to produce hypothetical structures for the SAPS domain, using other known helical repeat proteins such as importins, beta-catenin, PP2A scaffolding PR65 subunit and golgin p115. These models resemble one another in that they are made of multiple helices, but each has different helix segments, inter-helix loops, and positioning of the helices relative to one another. Because previous studies of PP2A binding to its helical scaffolding (PR65) or regulatory subunit alpha4 showed basic residues on the phosphatase catalytic subunit were required, and these presumably paired with acidic side chains in other subunit [27], we focused on acidic residues in PP6R3 as likely sites for contacts with PP6c. Reduced PP6 binding due to charge reversal mutations was taken as evidence for involvement of specific residues in subunit-subunit association. This analysis implicated E204, E205, E2659 and E262 as possible participants. Helical repeat models for SAPS domain based on beta-catenin (Figure 5B), or on golgin p115 (Figure 5C) position residues 204, 205, 259 and 262 proximal to one another, on the same side of the protein surface, in contrast to the 3D model based on PR65 A subunit (Figure 5A) that positions these loops on opposite sides. Therefore, results of our mutagenesis studies suggest either model B or C in Figure 5 for arrangement of the alpha helices in the SAPS domain.

The sensitivity of PP6R3 to proteolytic cleavage about 330-350 residues from the N terminus was noted with trypsin digestion of purified recombinant GST fusion protein, and with endogenous protease cleavage of various sized FLAG-tagged proteins expressed in intact cells, and with chymotrypsin digestion of FLAG-tagged PP6R3 immunoprecipitated from cells. In the 3D model based on the golgin p115 structure (Figure 5C) residues 320-340 are not in a helix, but instead predicted to be in an exposed surface loop (Figure 4), whereas in the beta-catenin structural model (Figure 5B) this segment is mostly in a helical conformation. There are two tandem Pro-Pro sequences at 323-324 and 338-339 likely to prevent alpha helix formation and the intervening sequence contains KKS and TWG as possible sites for trypsin and chymotrypsin. This putative surface loop and the juxtaposition of the 204, 205, 259 and 262 residues makes the golgin p115 structure, compose of ARM repeats, our favored model for the SAPS domain in PP6R3. We speculate that some structural element in the 350-513 sequence region, following this loop in the golgin p115 structure, constitutes a required contact site for binding to PP6c. The low sequence similarity between the golgin p115 and other ARM repeat proteins, despite ARM structures that are remarkably similar, tells us that the sampling of ARM, HEAT, ANK and PUM repeat proteins in the PDB database is still limited. This may account for why the SAPS domain sequence did not match any known structures.

Finally, binding of the PP6 catalytic subunit to the SAPS region of PP6R3 leaves the C terminal region of PP6R3 for interaction with the third subunit of the PP6 trimer, one of the ankyrin-repeat subunits (ARS), proteins known previously as Ankrd28, Ankrd44 and Ankrd52. This hypothetical arrangement is similar to, but different from, the actual 3D structures determined for PP2A heterotrimers by X-ray crystallography. Most obvious, the entire A subunit for PP2A is HEAT repeats, while no predictions suggest the same is true for PP6 subunits, where the SAPS domain only includes the N terminal half of the protein. The PR65 or A subunit scaffold for PP2A is an open arc of side-by-side helices in the AC dimer, but bends into a more closed horseshoe or letter "C" shape in the ABC trimer structures (see Fig. 5A). The B regulatory subunits or proteins such as small t antigen evidently induce the conformational change in the scaffold subunit [28]. Other evidence prior to the determination of the crystal structure indicated that small t antigen affected the conformation of the PP2A scaffold subunit [29]. The protease sensitivity of PP6R3 at ~330 residues from N terminus suggests that there may be a flexible or disordered and exposed junction. In PP6 trimers the subunit made of ankyrin repeats is expected to contact PP6c to alter substrate specificity while primarily tethered to the C terminal region of PP6R3, which alone is sufficient for stable association. The structures of known ankyrin-repeat proteins such as ankyrin, IκB, and MYPT1 show that neighboring helical repeats produce curvature in the overall structure [1,3]. Thus, we imagine that the PP6 catalytic and ARS subunits could come into proximity to one another by their mutual binding to PP6R3. Such a model, like what is seen with PP2A, would predict that substrate specificity and possible regulation of activity arises from interaction of the catalytic and regulatory subunits (ARS) brought together by tethering to a common scaffold (SAPS subunit).

Conclusion

In conclusion, the conserved SAPS domain in PP6R3 forms helical repeats that are probably organized similar to those in golgin p115. A core structure of about 350 N terminal residues is relatively protease-resistant and linked by a readily cleaved exposed segment to the rest of the SAPS region that is required for stable association with catalytic subunit. Charged residues in interhelical loops are used as primary contacts for specific recognition of PP6 and Sit4 catalytic subunits relative to other type-2A phosphatases.

Authors' Contributions

JG prepared antibodies and recombinant proteins, recorded the CD spectrum, produced the mutants, carried out pull-down assays and drafted portions of the manuscript, UD carried out sequence alignments and modelling, prepared Figures and drafted portions of the manuscript, JRE prepared recombinant proteins, carried out proteolysis analyses and drafted portions of the manuscript, DLB designed, organized and coordinated the studies, analyzed data, drafted portions of the manuscript and edited the final text. All authors read and approved the final manuscript.

Acknowledgements

This work was supported by grants from USPHS, NIH CA77584 and CA40042 (to DLB) and training grant T32 CA009109 (JRE). We thank Dr. WanChan Choi for help with the CD spectra and Dr. Zygmunt Derewenda for stimulating discussions.

References

1. Mosavi LK, Cammett TJ, Desrosiers DC, Peng ZY: The ankyrin repeat as molecular architecture for protein recognition. Protein Sci 2004, 13:1435–1448.

2. Andrade MA, Petosa C, O'Donoghue SI, Muller CW, Bork P: Comparison of ARM and HEAT protein repeats. J Mol Biol 2001, 309:1–18.

3. Li J, Mahajan A, Tsai MD: Ankyrin repeat: a unique motif mediating protein-protein interactions. Biochemistry 2006, 45:15168–15178.

4. Andrade MA, Bork P: HEAT repeats in the Huntington's disease protein. Nature Genetics 1995, 11:115–116.

5. Groves MR, Hanlon N, Turowski P, Hemmings BA, Barford D: The structure of the protein phosphatase 2A PR65/A subunit reveals the conformation of its 15 tandemly repeated HEAT motifs. Cell 1999, 96:99–110.

6. Xing Y, Xu Y, Chen Y, Jeffrey PD, Chao Y, Lin Z, Li Z, Strack S, Stock JB, Shi Y: Structure of protein phosphatase 2A core enzyme bound to tumor-inducing toxins. Cell 2006, 127:341–353.

7. Cho US, Xu W: Crystal structure of a protein phosphatase 2A heterotrimeric holoenzyme. Nature 2007, 445:53–57.

8. Xu Y, Xing Y, Chen Y, Chao Y, Lin Z, Fan E, Yu JW, Strack S, Jeffrey PD, Shi Y: Structure of the protein phosphatase 2A holoenzyme. Cell 2006, 127:1239–1251.

9. Xu Y, Chen Y, Zhang P, Jeffrey PD, Shi Y: Structure of a protein phosphatase 2A holoenzyme: insights into B55-mediated Tau dephosphorylation. Molecular cell 2008, 31:873–885.

10. Terrak M, Kerff F, Langsetmo K, Tao T, Dominguez R: Structural basis of protein phosphatase 1 regulation. Nature 2004, 429:780–784.

11. Sutton A, Immanuel D, Arndt KT: The SIT4 protein phosphatase functions in late G1 for progression into S phase. Molecular Cell Biology 1991, 11:2133–2148.

12. Bastians H, Ponstingl H: The novel human protein serine/threonine phosphatase 6 is a functional homologue of budding yeast Sit4p and fission yeast ppe1, which are involved in cell cycle regulation. Journal of Cell Science 1996, 109:2865–2874.

13. Luke MM, Della Seta F, Di Como CJ, Sugimoto H, Kobayashi R, Arndt KT: The SAP, a new family of proteins, associate and function positively with the SIT4 phosphatase. Molecular & Cellular Biology 1996, 16:2744–2755.

14. Stefansson B, Brautigan DL: Protein phosphatase 6 subunit with conserved Sit4-associated protein domain targets IkappaBepsilon.J Biol Chem 2006, 281:22624–22634.

15. Mi J, Dziegielewski J, Bolesta E, Brautigan DL, Larner JM: Activation of DNA-PK by ionizing radiation is mediated by protein phosphatase 6.PLoS ONE 2009, 4:e4395.

16. Morales-Johansson H, Puria R, Brautigan Dl, Cardenas ME: Human protein phosphatase PP6 regulatory subunits provide Sit4-dependent and rapamycin-sensitive Sap function in Saccharomyces cerevisiae. PLoS ONE 2009, 21;4(7):e6331.

17. Andrade MA, Perez-Iratxeta C, Ponting CP: Protein repeats: structures, functions, and evolution. Journal of structural biology 2001, 134:117–131.

18. Barton GJ: ALSCRIPT: a tool to format multiple sequence alignments. Protein engineering 1993, 6:37–40.

19. Ginalski K, Elofsson A, Fischer D, Rychlewski L: 3D-Jury: a simple approach to improve protein structure predictions. Bioinformatics (Oxford, England) 2003, 19:1015–1018.

20. Kelly SM, Price NC: The use of circular dichroism in the investigation of protein structure and function. Current protein & peptide science 2000, 1:349–384.

21. Striegl H, Roske Y, Kummel D, Heinemann U: Unusual armadillo fold in the human general vesicular transport factor p115. PLoS ONE 2009, 4:e4656.

22. Chen J, Peterson RT, Schreiber SL: α4 Associates with protein phosphatases 2A, 4 and 6.Biochimica et Biophysica Acta 1998, 247:827–832.

23. Prickett TD, Brautigan DL: The alpha-4 regulatory subunit exerts opposing allosteric effects on protein phosphatases PP6 and PP2A.J Biol Chem 2006, 281:30503–30511.

24. Nanahoshi M, Tsujishita Y, Tokunaga C, Inui S, Sakaguchi N, Hara K, Yoneza-wa K: Alpha4 protein as a common regulator of type 2A-related serine/threo-nine protein phosphatases. FEBS Lett 1999, 446:108–112.

25. McConnell JL, Gomez RJ, McCorvey LR, Law BK, Wadzinski BE: Identifica-tion of a PP2A-interacting protein that functions as a negative regulator of phosphatase activity in the ATM/ATR signaling pathway. Oncogene 2007, 26:6021–6030.

26. Prickett TD, Ninomiya-Tsuji J, Muratore TL, Shabanowitz J, Hunt DF, Brauti-gan DL: "Activation of TAK1-TAB1 Phosphorylation and Signaling to NF-kB by TAB4." J Biol Chem 2008, 283:19245–19254.

27. Prickett TD, Brautigan DL: Overlapping binding sites in protein phosphatase 2A for association with regulatory A and alpha-4 (mTap42) subunits. J Biol Chem 2004, 279:38912–38920.

28. Cho US, Morrone S, Sablina AA, Arroyo JD, Hahn WC, Xu W: Structural basis of PP2A inhibition by small t antigen. PLoS biology 2007, 5:e202.

29. Yang CS, Vitto MJ, Busby SA, Garcia BA, Kesler CT, Gioeli D, Shabanowitz J, Hunt DF, Rundell K, Brautigan DL, Paschal BM: Simian virus 40 small t antigen mediates conformation-dependent transfer of protein phosphatase 2A onto the androgen receptor. Mol Cell Biol 2005, 25:1298–1308.

Copyrights

Index

Printed and bound by CPI Group (UK) Ltd, Croydon, CR0 4YY

23/10/2024

01777682-0002